Continuum Mechanics Aspects of Geodynamics and Rock Fracture Mechanics

NATO ADVANCED STUDY INSTITUTES SERIES

Proceedings of the Advanced Study Institute Programme, which aims at the dissemination of advanced knowledge and the formation of contacts among scientists from different countries

The series is published by an international board of publishers in conjunction with NATO Scientific Affairs Division

A	Life Sciences	Plenum Publishing Corporation
B	Physics	London and New York
C	Mathematical and Physical Sciences	D. Reidel Publishing Company Dordrecht and Boston
D	Behavioral and Social Sciences	Sijthoff International Publishing Company Leiden
E	Applied Sciences	Noordhoff International Publishing Leiden

Series C – Mathematical and Physical Sciences

Volume 12 – Continuum Mechanics Aspects of Geodynamics and Rock Fracture Mechanics

Continuum Mechanics Aspects of Geodynamics and Rock Fracture Mechanics

Proceedings of the NATO Advanced Study Institute held in Reykjavik, Iceland, 11–20 August, 1974

edited by

P. THOFT-CHRISTENSEN

Aalborg Universitetscenter Matematik, Danmarks Ingeniørakademi, Aalborg, Danmark

D. Reidel Publishing Company

Dordrecht-Holland / Boston-U.S.A.

Published in cooperation with NATO Scientific Affairs Division

ISBN-13: 978-94-010-2270-5 e-ISBN-13: 978-94-010-2268-2
DOI: 10.1007/978-94-010-2268-2

Published by D. Reidel Publishing Company
P.O. Box 17, Dordrecht, Holland

Sold and distributed in the U.S.A., Canada, and Mexico
by D. Reidel Publishing Company, Inc.
306 Dartmouth Street, Boston, Mass. 02116, U.S.A.

Softcover reprint of the hardcover 1st edition 1974

CONTENTS

PREFACE

During a NATO Advanced Study Institute in Izmir, Turkey, July 1973 on Modern Developments in Engineering Seismology and Earthquake Engineering it emerged that a debate on Continuum Mechanics Aspects of Geodynamics and Rock Fracture Mechanics would be very welcome. Therefore, it was decided to seek NATO sponsorship for an Advanced Study Institute on this subject.

The purpose of the new Advanced Study Institute was to provide a link between mechanics of continuum media and geodynamics. By bringing together a group of leading scientists from the above two fields and participants actively engaged in research and applications in the same fields, it was believed that fruitful discussions could emerge to facilitate an exchange of knowledge, experience and newly-conceived ideas.

The Institute aimed primarily at the solution of such problems as connected with the study of stress and strain conditions in the Earth, generic causes of earthquakes, energy release and focal mechanism and seismic wave propagation introducing modern methods of continuum and rock fracture mechanics. Secondly to inspire scientists working in continuum mechanics to open new avenues of research connected with the above problems, and seismologists to adapt modern, advanced methods of continuum and rock fracture mechanics to their work.

Geophysics is one of the most exciting subfields of physics. The main reason for this is perhaps that geophysics is a research area, that generally cannot be controlled by the observer. Further this field is very fascinating because it relate directly to the relationship between man and nature. Finally a characteristic aspect of this field is its problemoriented nature and that scientists with very different backgrounds in physics, mathematics, engineering and so on here work together and are forced to look into each others problems.

The traditional research in geophysics used perhaps to be based more on technical and descriptive methods rather than on fundamental understanding of the natural phenomena. But this seems now to have changed completely. Geophysics became a major area of research after the first World War due to the oil and mining industry, but after the second World War the theory of seafloor spreading has increased the importance of geophysics so drasticly that one can talk about a revolution in geophysics. A completely new picture of the earth's crust with large plates floating on the underlying mantle is developed. This model has open up the possibility of getting a reliabel explanation of such phenomena as continental drift, sea-floor spreading, mountain building, seismic zones and volcanics activity. Prediction of the occurence of earthquakes is perhaps a possibility in few years and it will some days perhaps even be possible to prevent earthquakes by injecting fluids to relieve strain along rock fractures.

The central idea in the theory of plate tectonics is the excistens of a rigid upper layer, which has a considerable strength and is roughly 100 kilometers thick. This layer rests or floats on a second layer, which has essentially no strength and a thickness of several hundred kilometers. The second layer is assumed to offer practically no resistance to the horizontal movement of the upper layer. The upper layer is divided into large plates which are bounded by the ocean ridges and by certain faults.

From the point of view of continuum mechanics the theory of plate tectonics is of great interest and raise a lot of interesting problems. How is the force system responsible for the movements of the plates arranged? Is the movement due to differences in temperatures under the oceans and the continents? Is it possible by considering the earth as a mechanical model to calculate in details the motion of the plates, the occurence of earthquakes etc.?

All aspects of modern continuum mechanics are needed to answer such questions. Can the plates be considered rigid, elastic, plastic or viscoelastic or do we need a more sophisticated theory? Are the plates homogeneous and isotropic? Is it possible to obtain good solutions with regard to wave propagation in the earth?

A new period began in geophysics with the theory of plate tectonics twenty years ago. In continuum mechanics a new period began in 1945. The new period is characterized by work on non-linear phenomena, particularly in the case of large deformations. On a sound basis the well-known theories have been supplemented with new theories able to take into consideration nearly all situations.

This new period in continuum mechanics can also be characterized by the fact that continuum mechanics to day is based on more general principles than it used to be. But, unfortunately the physics behind the new theories often cannot follow up with the mathematical manipulations.

Therefore solving real problems in geophysics perhaps may lead to new improved theories of great practical value. The problems are there - the challenge is great.

P. Thoft-Christensen.

SCIENTIFIC DIRECTORS

Thoft-Christensen, P.	Aalborg University Center Danmarksgade 19 9000 Aalborg Denmark
Solnes, J.	University of Iceland Reykjavik Iceland

LECTURERS

Bullen, K. E.	University of Sydney Sydney, N.S.W. 2006 Australia
Erdogan, F.	Lehigh University Bethlehem Pennsylvania 18015 USA
Eringen, A.C.	Princeton University E-307 Engineering Quadrangle Princeton, N.J. 08540 USA
Kröner, E.	Inst. für Theor. und Angew. Physik Universität Stuttgart 7 Stuttgart 1 W. Germany
Palmason, G.	Lamont-Doherty Geological Observatory of Columbia University Palisades, N.Y. 10964 USA
Rivlin, R.S.	Lehigh University Bethlehem Pennsylvania 18015 USA

Rogers, T.G.	Dept. of Theoretical Mechanics University of Nottingham University Park Nottingham NG 7 2RD England
Teisseyre, E.	Inst. of Geophysics Polish Academy of Science Pasteura 3 00-973 Warsaw Poland
Tryggvason, E.	Dept. of Earth Sciences University of Tulsa 600 South College Tulsa, Oklahoma 74104 USA

PARTICIPANTS

Armand, J.-L.	Dept. of Mechanics Ecole Polytechnique 12 Avenue Boudon 75016 Paris France
Atluri, S.	Georgia Institute of Technology 225 North Avenue, N.W. Atlanta, Georgia 30332 USA
Batterman, S.C.	University of Pennsylvania 111 Towne Building Philadelphia 19174 USA
Bjørnsson, S.	University of Iceland Reykjavik Iceland
Boulanger, P.	Université Libre de Bruxelles Departement de Mathematique Avenue F.-D. Roosevelt, 50 1050 Bruxelles Belgium
Byskov, E.	Danmarks Tekniske Højskole Bygning 118 2800 Lyngby Denmark

Caiado, V.	Geophysical Institute of Lisbon University Rua da Escola Politecnica Lisbon Portugal
Cetincelik, M.	Dept. of Earthquake Engineering P.O. Box 400 Kizilay Ankara Turkey
Drescher, A.	Inst. of Fund. Tech. Res. Polish Academy of Sciences Swietokrzyska 21 00-049 Warsaw Poland
Einarson, T.	University of Iceland Reykjavik Iceland
Finn, W.D.L.	Faculty of Applied Science University of British Columbia Vancouver, B.C. Canada
Gunnlaugsson, G.A.	University of Iceland Reykjavik Iceland
Hanagud, S.	Georgia Institute of Technology 225 North Avenue, N.W. Atlanta, Georgia 30332 USA
Harder, N.A.	Aalborg University Center Danmarksgade 19 9000 Aalborg Denmark
Jacobsen, M.	Aalborg University Center Danmarksgade 19 9000 Aalborg Denmark
Jensen, Aa.P.	Danmarks Ingeniørakademi Bygning 373 2800 Lyngby Denmark

Karaesmen, E.	Dept. of Civil Engineering Black Sea Technical University Trabzon Turkey
Karlsson, T.	University of Iceland Reykjavik Iceland
Krenk, S.	Danmarks Tekniske Højskole Bygning 118 2800 Lyngby Denmark
Kusznir, N.J.	Dept. of Geological Science University of Durham Durham England
Neugebauer, H.	Johan Wolfgang Goethe-Universität Feldbergstrasse 47 6 Frankfurt a.M. 1 W.Germany
Ramstad, L.J.	Inst. for Statikk NTH Trondheim Norway
Rathkjen, A.	Aalborg University Center Danmarksgade 19 9000 Aalborg Denmark
Sabina, F.J.	Instituto de Geofisica Torre de Ciencias Ciudad Universitaria Mexico 20, D.F.
Sandbye, P.	Danmarks Ingeniørakademi Bygning 373 2800 Lyngby Denmark
Sawyers, K.N.	Lehigh University Bethlehem Pennsylvania 18015 USA

Seide, P.	University of Southern California Dept. of Civil Engineering Los Angeles, Calif. 90007 USA
Selvadurai, A.P.S.	Dept. of Civil Engineering University of Aston Gosta Green Birmingham B4 7ET England
Sigbjørnsson, R.	University of Trondheim NTH Trondheim Norway
Steketee, J.A.	Delft University of Technology Dept. of Aeronautical Eng. Kluyverweg 1 Delft Netherlands
Thomsen, L.	Dept. of Geological Sciences State University of N.Y. Binghamton, N.Y. 13901 USA
Wilson, R.C.	University of Utah Salt Lake City Utah 84112 USA
Withers, R.J.	Physics Department University of Alberta Edmonton Canada
Woodhouse, J.H.	Dept. of Applied Mathematics and Theoretical Physics University of Cambridge Silver Street Cambridge CB3 9EW England

ASPECTS OF EARTHQUAKE ENERGY

K. E. Bullen

c/o Department of Applied Mathematics,
University of Sydney, Australia

ABSTRACT. Some aspects of the energy in seismic waves are discussed, with special reference to the problem of estimating the total energy released in earthquakes. A calculation is presented connecting the energy of a large earthquake with the size of the region in which significant deviatoric strain has accumulated prior to the earthquake.

1. EXPRESSIONS FOR ENERGY IN SIMPLE ONE-DIMENSIONAL WAVE TRANSMISSION

Let v be the velocity of a train of waves advancing along the x-axis in a uniform deformable medium. The displacement u may be represented at time t by the form

$$u = f(x - vt)$$

$$= \sum A_r \cos\{2\pi(x/\lambda_r - t/\tau_r) + \varepsilon_r\}, \quad (1.1)$$

(the summation may need to be replaced by an integral), where A_r denotes the amplitude, λ_r the wave length, and τ_r the period of a sinoidal constituent.

Let W be the mean energy in the wave motion, per unit volume of the medium. Half this energy is kinetic and half potential (see ref. 1, §3.3.6). Thus W is twice the mean kinetic energy per unit volume.

In a portion of the medium of length b (say) parallel to the x-axis, unit cross-sectional area and density ρ, the kinetic energy is

Thoft-Christensen (ed.), Continuum Mechanics Aspects of Geodynamics and Rock Fracture Mechanics, 1-12.

$$\int_0^b \tfrac{1}{2}\rho\,(\partial u/\partial t)^2 dx. \qquad (1.2)$$

To obtain W, we have to divide (1.2) by b and let b → ∞ (in order to get the required mean value) and then double. On substituting from (1.1) and reducing, we obtain

$$W = \sum 2\pi^2 A_r^{\,2} \tau_r^{\,-2} \rho. \qquad (1.3)$$

In particular, for a purely sinoidal wave train, we have, dropping subscripts,

$$W = 2\pi^2 A^2 \tau^{-2} \rho. \qquad (1.4)$$

The mean energy W' in a portion of the medium of length λ and unit cross-sectional area is

$$W' = 2\pi^2 A^2 \lambda \tau^{-2} \rho. \qquad (1.5)$$

2. EARLY METHODS OF ESTIMATING SEISMIC WAVE ENERGY

2.1 Preliminary remarks

It would be theoretically possible, using formulae based on (1.3)-(1.5), to estimate to closer precision the wave energy, E say, released in an earthquake if sufficiently well determined measurements could be made on seismogram records taken at a sufficient number of suitably distributed stations on the Earth's surface. In practice, many difficulties make the task formidable and, moreover, complicated by greater or less uncertainties at several stages of the process. Following is an outline of some early attempts at approximations.

2.2 Use of records at nearby stations

With some earthquakes, useful estimates of E can be derived from records of SH waves at nearby stations. A sizable fraction of the total bodily wave energy can usually be expected to be in SH waves, the treatment of which is much simpler than for P and SV since SH are reflected and refracted only into SH waves.

For the Jersey earthquake of 1926 July 20, Jeffreys [2] noted that SH bodily waves that had travelled through a near-surface crustal layer were comparatively large at epicentral distances up to 500 km. He assumed that the energy in these waves travelled out from the focal region (presumed to lie inside the layer) with a cylindrical wave front inside the layer, and that this energy approximated to E. Let ρ and H be the density and thickness of the layer, Δ the angular epicentral distance of a recording station Q, and let r_0 be the Earth's radius. Treating the waves as a sinoidal train of amplitude A, period τ and total length L, and using (1.5), he arrived at

$$E \sim 4\pi^3\rho(r_0\sin\Delta)HLA^2\tau^{-2}$$

$$= \pi\rho(r_0\sin\Delta)HLv_m^2, \qquad (2.1)$$

where v_m is the maximum velocity of the ground motion.

From records at each of several single stations, Jeffreys used (2.1) to estimate that $E \sim 10^{12}$ J for the Jersey earthquake.

Since bodily waves are not usually closely sinoidal, a slightly more accurate formula would be

$$E \sim 4\pi^3\rho(r_0\sin\Delta)H\beta\int A^2\tau^{-2}dt, \qquad (2.2)$$

where β is the wave velocity, and the integration is over the arriving SH train.

2.3 Estimation of bodily wave energy from records at distant stations

Assuming a spherically symmetrical issue of bodily waves from the focal region F, and treating the Earth as uniform, Galitzin derived a formula which, as later modified, is equivalent to

$$E \sim 4\pi^3\rho\beta\{2r_0\sin(\Delta/2)\}^2\int A^2\tau^{-2}d\tau, \qquad (2.3)$$

where A and τ relate to the SH waves recorded at a station Q. With (2.3), F is assumed to be at the Earth's surface, the wave energy therefore issuing downward from F. With SH waves, the calculation is assisted by the theoretical result that the amplitudes of the waves emerging at Q are half those of the surface ground movement. (The corresponding results for P and SV are more complicated.)

When the focal depth h is appreciable (the wave energy now issuing upward as well as downward from F), (2.3) needs to be replaced by

$$E \sim 8\pi^3\rho\beta\{h^2 + 4r_0(r_0 - h)\sin^2(\Delta/2)\}\int A^2\tau^{-2}dt. \qquad (2.4)$$

If Q is at the epicentre, (2.4) reduces to

$$E \sim 8\pi^3\rho\beta h^2\int A^2\tau^{-2}dt. \qquad (2.5)$$

A formula equivalent to (2.5) was made the basis of a method of Gutenberg and Richter for estimating E.

2.4 Estimation of energy in surface waves

Jeffreys used simple Rayleigh wave theory to estimate the order of magnitude of the energy in P-SV surface waves. He arrived at a formula of the form (2.2), with H replaced by 1.1 λ, where λ is the wave length and A is the horizontal component of the surface ground motion. (For details, see ref. 1, §15.1.3.)

2.5 Application

Depending on the characteristics of an earthquake (magnitude, focal depth, relative proportions of energy in bodily and surface waves), formulae based on those in §§2.2-2.4 have been much used in efforts to estimate E. This applies in particular in the pioneering work of Gutenberg and Richter [3] on the Earth's seismicity.

3. STEPS TOWARDS IMPROVED PRECISION

The simplifications in §2 are of course fairly drastic. A step towards improved precision is indicated below.

3.1 Taking account of continuous variation of velocity with depth

Consider (say) P waves issuing symmetrically from a focal region F and travelling to points Q at the Earth's surface with velocity α which depends on the distance r from the centre of the Earth, here assumed spherically symmetrical. Assume for the present that the waves are continuously refracted between F and Q and encounter no internal surfaces of discontinuity. Let Δ be the angular epicentral distance of Q, and let $\eta = r/\alpha$. At any point of a ray, let e be the angle between the ray and the level surface through the point. For the ray FQ, let $e = e_1, e_0$ at F, Q, respectively.

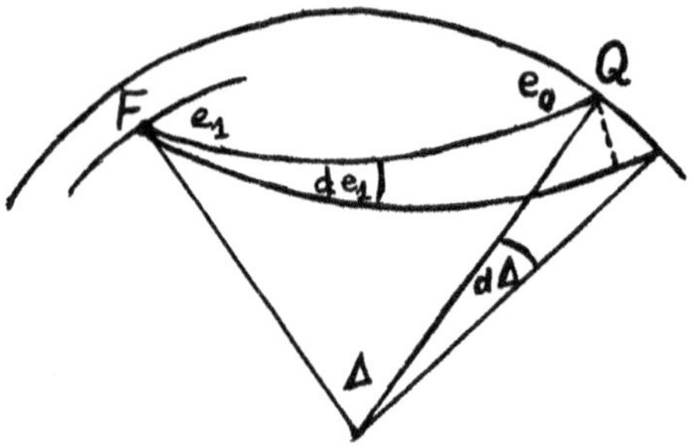

Let I be the energy, per unit solid angle, in the waves in question as they issue from F. Then the energy dI being transmitted through the volume bounded by rays for which $e = e_1$, $e_1 + de_1$ is given by

$$dI = 2\pi I |de_1| \cos e_1. \tag{3.1}$$

The area of the Earth's surface at which this energy emerges is $2\pi r_0^2 |d\Delta| \sin\Delta$. The corresponding area dA of the emerging wave front is

$$dA = 2\pi r_0^2 \sin\Delta \sin e_0 |d\Delta|. \tag{3.2}$$

Hence, neglecting all energy dissipation en route, the energy $U(\Delta)$ per unit area of wave front emerging at Q is

$$U(\Delta) = \frac{dI}{dA} = \frac{I \cos e_1}{r_0^2 \sin\Delta \sin e_0} \left|\frac{de_1}{d\Delta}\right|. \tag{3.3}$$

Let T be the travel time along the ray FQ. Then, by standard seismic ray theory,

$$\eta_1 \cos e_1 = \eta_0 \cos e_0 = dT/d\Delta, \tag{3.4}$$

whence (see ref. 1, §8.1)

$$U(\Delta) = \frac{I\eta_0}{r_0^2 \eta_1 \sin\Delta} \left(\eta_1^2 \tan^2 e_0 - \eta_0^2 \sin^2 e_0\right)^{-\frac{1}{2}} \frac{d^2T}{d\Delta^2}. \tag{3.5}$$

3.2 Limitations of the formula

The formula (3.5), though superior to (2.4) through taking account of variation of α with r, still ignores several complications that are significant in practice.

The formula is inadequate for waves which have encountered one or more surfaces of discontinuity between F and Q. Incident P waves may be converted at such a surface into P and SV reflected and P and SV refracted waves. For any one of the four sets of converted waves, an energy 'transmission factor' has to be applied to formulae of the type (3.5). Such factors vary substantially with the angle of incidence at the discontinuity surface and are subject to uncertainties, which may be considerable, as to the character and location of the discontinuity. (For some details on transmission factors, see ref. 1, chapters 5 and 8.)

Sufficiently rapid changes of property inside the Earth may also cause conversion of energy. Sometimes, depending on the wave lengths involved, a rapid change may be treated as a discontinuity. (A mathematical discontinuity is of course only a mathematical model concept.) Where a rapid change cannot be treated as a discontinuity, complex analysis may be required. For an indication of the type of mathematics needed, see refs. [4], [5], [6] and [7, §8.4].

Through (3.5), modified if necessary by the inclusion of transmission factors or their equivalent, a fair estimate of the total seismic wave energy can sometimes be made from data at a limited number of stations. In practice, data from a wide-spread distribution of stations is likely to be required because of asymmetry at the focus. The observational uncertainties, as well as uncertainties on the distributions of α and β with r, contribute to the uncertainty of the estimated energy.

Account has also to be taken of departures from spherical symmetry in the Earth. Departures associated with the ellipticities of surfaces of constant velocity in the Earth have only minor effects and could if needed be readily allowed for. But departures due to lateral variations of wave velocity, especially in the crust, have more serious effects. There is no ready way of dealing with these except by long trial and error, and slow accumulation of evidence on the three-dimensional velocity distributions. Limitations of this evidence add further to the uncertainties.

Energy losses also occur through scattering (see e.g. ref. 8) and departures from perfect elasticity. Båth [9] estimated that, with bodily waves from shallow-focus earthquakes, the losses inside the crust (including losses connected with lateral variations) may involve an energy 'extinction factor' as high as 20. The factor is greater for S than P waves, and Båth regarded the high extinction of short-period bodily waves near the focal region as one of the more serious sources of uncertainty in estimating earthquake energy. He also estimated that the total extinction during transmission inside the mantle is 10-15 per cent of that inside the crust. For waves travelling long distances D, attenuation factors of the form e^{kD} are sometimes introduced. (See again ref. 9.)

The energy in surface waves is not taken into account in (3.5). If this energy is not independently estimated (see e.g. §2.4), a further factor has to be applied to allow for it. The factor varies from earthquake to earthquake and is specially sensitive to the focal depth.

The summary is that, although formulae of the type (3.5) have led to some increase in precision, it is not yet possible to estimate the energy of an earthquake within a factor of at least 2: usually the uncertainty factor is appreciably greater than 2.

For detail of some further approaches, see Knopoff [10], Belotelov, Kondorskaya and Savarensky [11], DeNoyer [12] and Randall [13].

4. ESTIMATION OF ENERGY FROM STRAIN MEASUREMENTS

A wholly different approach to the problem of earthquake energy is through the application of geodetic data to estimating the strain energy in the vicinity of the focal region before and after an earthquake. The strain measurements are made in the vicinity of geological faults at the surface, and assumptions are made on the faulting and strain below. Examples of earthquake energy calculations made in this way are those of Byerly and DeNoyer [14]. They gave for the San Francisco earthquake of 1906 April 18, $E = 0.9 \times 10^{16}$ J; the Imperial Valley earthquake of 1940, 0.96×10^{15} J; and the Nevada earthquake of 1954 December 16, $1\text{-}1.5 \times 10^{15}$ J.

5. EARTHQUAKE ENERGY AND MAGNITUDE

The discussions in §§2,3 make it evident that the close determination of earthquake energy must have considerable recourse to empirical methods. Details of these methods are closely linked with estimations of earthquake magnitudes. The present section briefly outlines some of the principal results.

The first magnitude scale [15] defined the magnitude M in terms of the maximum amplitude traced by a standard seismograph (free period 0.8 s; statical magnification 2800; damping coefficient 0.8) at an epicentral distance of 100 km. Empirical tables were set up with a view to reducing observations taken at other distances and on other types of seismographs to results corresponding to Richter's standard conditions. Originally, only shallow-focus earthquakes were considered, but the tables were later extended to allow for significant focal depth. Subsequently, various modifications were made to the magnitude scale itself. On the latest scale, the largest earthquakes have $M = 8.9$.

In a long series of papers, Gutenberg and Richter sought to connect M with the earthquake energy E by the form

$$aM = \log_{10}(E/E_0), \tag{5.1}$$

bringing vast quantities of empirical data to bear. A recent revision by Båth [16] gave

$$\log_{10}E = 5.24 + 1.44\,M, \tag{5.2}$$

where E is in joules; this gives $E \approx 10^{18}$ J for $M = 8.9$, and $E \approx 1.7 \times 10^5$ J for a zero magnitude earthquake (conventionally presumed to correspond to the smallest recorded earthquakes).

Formulae of the type (5.2), along with other observational evidence, have been applied with much success to estimate many aspects of earthquake energy release. For example, Gutenberg [17] estimated that the total annual release of earthquake energy is 10^{18} J, corresponding to a rate of work of 10^{7}-10^{8} kW. This is about 10^{-3} times the rate of heat escape from the Earth's interior. (It has sometimes been suggested that the Earth acts as a heat engine converting a small fraction of the escaping heat into strain energy.) It is interesting that the energy in a major hurricane is of the same order as that in an extreme earthquake.

Eighty per cent of the total energy in all earthquakes comes from those for which $E = 10^{16}$-10^{18} J. A table by Båth [16] gives the following percentages of earthquake energy release in different geographic regions: North America (including Alaska), 10; South America, 16; Southwest Pacific and Philippines, 26; Ryukyu-Japan, 16; Kurile Islands, Kamchatka and Aleutians, 9; Central Asia, 17; Indian and Atlantic Oceans, 6.

Båth and Duda [18], assuming that the volume V (m^3) of the strained region prior to a large earthquake is about equal to the volume encompassing the aftershocks, derived empirically

$$\log_{10} V = 3.58 + 1.47\ M. \qquad (5.3)$$

The formula (5.3) has some interest in connection with the calculations in §7.

6. NUCLEAR EXPLOSIONS AND EARTHQUAKE ENERGY

Since nuclear explosions are in certain respects of the nature of controlled earthquakes, with knowledge available of the total released energy, the source location and time of origin, there is the theoretical possibility of using them to derive information on the energy released in natural earthquakes. There are, however, several practical difficulties in obtaining useful results in this way.

On distant records of underground nuclear explosions, S and surface waves are often weak. For this reason alone, the formula (5.2) may give $\log_{10} E$ too great by unity or more if M is estimated by the usual procedures for natural earthquakes.

More important is the size and variability of the seismic efficiency f (the ratio of the seismic wave energy caused by the explosion to the total energy released). Average values of f are as follow: explosions in the atmosphere at altitudes 1-10 km, $O(10^{-5})$; at the Earth's surface, $O(10^{-4})$; 300 m underground, $O(10^{-3})$; 30 m underwater, $O(5\times10^{-3})$; 300 m underwater, $O(10^{-2})$. The values vary widely with the source conditions: for an explosion inside a large underground cavity, f may be less than 10^{-2}

that for a well-tamped explosion at the same depth. For further details, see ref. 1 (chapter 16) and ref. 16 (chapter 11).

7. EARTHQUAKE ENERGY AND EXTENT OF STRAINED REGION

The extent of the strained region prior to a large earthquake can be assessed from early calculations of the writer [19,20] given below. Several investigators, e.g. Tsuboi [21], independently arrived later at similar results derived on a somewhat narrower basis.

7.1 Preliminaries on stress-strain relations and strain energy

For present purposes, it is sufficient to assume perfect elasticity, isotropy and linear strain theory. Then the set of stress-strain relations may be written as

$$p_{ij} = (k - 2\mu/3)\theta\delta_{ij} + 2\mu e_{ij}, \tag{7.1}$$

where the p_{ij} and e_{ij} are the components of ordinary stress and strain, θ $(= \Sigma e_{kk})$ is the dilatation, δ_{ij} is the Kronecker delta, k is the incompressibility and μ the rigidity. (It is preferable to use k and μ, which have immediate physical significance, rather than pairs such as the Lamé parameters λ and μ.)

The deviatoric stress and strain components P_{ij} and E_{ij} are defined by

$$P_{ij} = p_{ij} - \tfrac{1}{3}\Sigma p_{kk}\delta_{ij}, \tag{7.2}$$

$$E_{ij} = e_{ij} - \tfrac{1}{3}\Sigma e_{kk}\delta_{ij}. \tag{7.3}$$

(All summations are from 1 to 3 and are with respect to repeated subscripts.) By (7.2) and (7.3), the stress-strain relations (7.1) may be re-written as

$$p_{kk} = 3k\theta; \quad P_{ij} = 2\mu E_{ij}. \tag{7.4}$$

The relations (7.4) have the important advantage that the physically significant parameters k and μ appear in separate equations.

The strain energy W per unit volume at a point of a strained body is given [1, §2.3.5] by

$$W = \tfrac{1}{2}k\theta^2 + \mu(\Sigma\Sigma e_{ij}{}^2 - \tfrac{1}{3}\theta^2). \tag{7.5}$$

By (7.2) and (7.3), this becomes

$$W = \tfrac{1}{2}k\theta^2 + \mu\Sigma\Sigma E_{ij}{}^2. \tag{7.6}$$

The two terms on the right side of (7.6) give the compressional and the deviatoric strain energy per unit volume, respectively.

7.2 Strength

Let p_i (i = 1,2,3) be the principal stresses at a point Q of a stressed body and let primes indicate values of stress components at the stage when, under increasing stress, flow or fracture starts to occur at Q. Let $p_1' \geq p_2' \geq p_3'$. The strength at Q is commonly defined in terms of the values of certain functions of the p_i'. Two different functions have been used: the stress-difference $p_1' - p_3'$; and the Mises function S, where

$$S^2 = (p_1' - p_3')^2 + (p_2' - p_1')^2 + (p_3' - p_2')^2. \tag{7.7}$$

From (7.2) and (7.3), it can be deduced that

$$S^2 = 3\Sigma\Sigma(P_{ij}')^2. \tag{7.8}$$

The strength sets an upper bound to the possible value of $p_1 - p_3$, or of $\sqrt{(3\Sigma\Sigma(P_{ij})^2)}$ on the two definitions, respectively.

By simple algebra, it can be shown that S lies between 1.22 and 1.42 times the stress-difference. Since only orders of magnitude of S are involved in geophysical applications, it does not matter which definition is used. The Mises strength is used below.

7.3 Connection between energy, strength and rigidity

Just before a large earthquake, let V be the volume of the region R (surrounding the focus) inside which there is significant deviatoric strain. At any point of R,

$$3\Sigma\Sigma P_{ij}^2 = \alpha S^2, \tag{7.9}$$

where $0 \leq \alpha \leq 1$; for $3\Sigma\Sigma P_{ij}^2 \leq S^2$.

Corresponding to (7.6), the total strain energy E_s inside R is equal to βE_d, where β (>1) is the ratio of E_s to the deviatoric strain energy E_d, and

$$E_d = \iiint \mu\Sigma\Sigma E_{ij}^2 d\tau, \tag{7.10}$$

the integration being through V.

Let E be the energy released in the form of seismic waves, and write $E = \beta\gamma E_d$. It is to be expected that $\beta \sim O(1-2)$ and $\gamma \leq 0.5$. For the purpose of an order of magnitude calculation it is appropriate to take $\beta\gamma = 0.5$. Then

$$E \approx 0.5\iiint \mu\Sigma\Sigma E_{ij}^{2} d\tau. \tag{7.11}$$

For simplicity, μ and S will be treated as constant throughout V. Then, using (7.4) and (7.9), we have

$$\begin{aligned} 24\mu E &\approx \iiint 3\Sigma\Sigma P_{ij}^{2} d\tau \\ &= S^2\iiint \alpha d\tau \\ &= S^2 V_0, \text{ say,} \end{aligned} \tag{7.12}$$

where V_0 would be the volume of R if the material had been about to fracture or flow at every point of R.

7.4 Implications of equation (7.12)

Inside the range of depth at which earthquakes originate, the rigidity μ is known to lie (in effect) between about 0.4 and 1.5×10^{11} N/m^2. For the largest earthquakes, $E \approx 10^{18}$ J (§5). Thus (7.12) gives

$$S^2 V_0 = O(10^{30}\ N^2/m), \tag{7.13}$$

whence S and V_0 must both be considerable in a large earthquake.

Laboratory evidence indicates that for rocks in the outer part of the Earth, $S \leq O(10^8\ N/m^2)$. Thus (7.13) gives

$$V_0 \geq O(10^{14}\ m^3) = O(10^5\ km^3). \tag{7.14}$$

This result seemed surprising when first derived, though it has since been amply confirmed. It implies that the strained region would occupy a volume at least equal to the volume of a sphere of 50 km diameter, even if the material were about to fracture or flow throughout this volume. Since the material would actually be well short of this condition throughout most of R, V must be considerably greater than V_0, perhaps exceeding the volume of a sphere of diameter 100 km. Furthermore, it is improbable that R would be spherical. Hence one or two of the dimensions of R would probably be well in excess of 100 km, thus tending towards the order of the Earth's radius.

The result (7.14) may be compared with the result obtained using the empirical formula (5.3) which, for an earthquake of magnitude 8.9, would give $V \approx 5 \times 10^{16}$ m^3. The result (7.14) also played an important rôle in the reduction made by Gutenberg and Richter from 10^{20} to 10^{18} J as their estimate of the energy in an extreme earthquake. In addition, it showed that the strength S cannot be much less than 10^8 N/m^2 where a large earthquake occurs.

The finding that one or more of the dimensions of V could be so large provided some indirect support for the notion of possible causal connections between globally wide-spaced large earthquakes. Benioff [22] had suggested that earthquakes for which M > 8.0 may not be entirely independent events, but are related to a global stress system.

REFERENCES

1. K. E. Bullen, Introduction to the Theory of Seismology, University Press, Cambridge, 3rd ed, 1965.
2. H. Jeffreys, Mon. Not. R. Astr. Soc.* 1, 483, 1927.
3. B. Gutenberg and C. F. Richter, Seismicity of the Earth and associated phenomena, University Press, Princeton, 2nd ed, 1954.
4. J. G. J. Scholte, Kon. Ned. Meteorol. Inst. 65, 1, 1957.
5. L. Cagniard, Réflexion et Réfraction des Ondes Séismiques Progressives, Gauthier-Villars, Paris, 1939. (English trans. by E. A. Flinn and C. H. Dix, McGraw-Hill, New York, 1962.)
6. B. L. v. d. Waerden, Reflection and Refraction of Seismic Waves, Shell Development Company, 54 pp., 1957.
7. M. Båth, Mathematical Aspects of Seismology, Elsevier, Amsterdam, 1968.
8. R. A. W. Haddon and J. R. Cleary, Phys. Earth Planet. Interiors 8, 211, 1974.
9. M. Båth, in Contributions in Geophysics, Pergamon, London, pp. 1-16, 1958.
10. L. Knopoff, Geophys. J., Roy. Astr. Soc., 1, 44, 1958.
11. V. L. Belotelov, N. V. Kondorskaya and E. T. Savarensky, Ann. di Geofis. 14, 57, 1961.
12. J. DeNoyer, Bull. Seismol. Soc. Amer. 48, 353, 1958; and 49, 1, 1959.
13. M. J. Randall, Bull. Seismol. Soc. Amer. 63, 1133, 1963.
14. P. Byerly and J. DeNoyer, in Contributions in Geophysics, Pergamon, London, pp. 17-35, 1958.
15. C. F. Richter, Bull. Seismol. Soc. Amer. 25, 1, 1935.
16. M. Båth, Introduction to Seismology, Birkhäuser Verlag, Basel, 1973.
17. B. Gutenberg, Quart. J. Geol. Soc. Lond. 112, 1, 1956.
18. M. Båth and S. J. Duda, Ann. di Geofis. 17, 353, 1964.
19. K. E. Bullen, Trans. Amer. Geophys. Un. 34, 107, 1953.
20. K. E. Bullen, Bull. Seismol. Soc. Amer. 45, 43, 1955.
21. C. Tsuboi, J. Phys. Earth 4, 63, 1956.
22. H. Benioff, Bull. Geol. Soc. Amer. 65, 385, 1954.

---

* Geophys. Suppl.

CONSTRUCTION OF EARTH MODELS

K. E. Bullen

c/o Department of Applied Mathematics,
University of Sydney, Australia

ABSTRACT. An outline is given of methods used to construct model distributions of the density, pressure, incompressibility, rigidity, gravitational intensity and P and S seismic velocities in the Earth's interior. Spherical symmetry is assumed. Reference is made to the problem of formulating a standard Earth model. A table giving values of various properties of the Earth's interior at selected depths is included.

1. INTRODUCTION

The Earth models to be discussed in this paper give model distributions of the density ρ, pressure p, incompressibility k, rigidity μ, gravitational intensity g, and P and S seismic velocities α and β in the Earth's interior.

An ultimate aspiration is to derive reliable values of these properties at points of the interior whose positions are specified in terms of three space variables. Consideration will, however, here be limited to spherically symmetrical models, the properties being thus expressed in terms of the distance r from the centre O, or depth z below the surface. Data are available [1] from which models taking account of the ellipticities of surfaces of constant density within the Earth can be readily derived; but fine detail taking account of other deviations from spherical symmetry is not adequately available as yet. Thus the models give, in some sense, laterally averaged values of properties. Incidentally, non-symmetrical models would (apart from ellipticity) involve further complications; e.g. the stress would not be adequately represented by the single parameter p - in solid regions there would be non-zero deviatoric stresses.

Thoft-Christensen (ed.), Continuum Mechanics Aspects of Geodynamics and Rock Fracture Mechanics, 13-21.

The sets of observational evidence brought to bear in constructing Earth models include:

(i) Data on the Earth's mean radius R, mass M and mean moment of inertia I. The uncertainties of these data are now sufficiently small, compared with other uncertainties, to be neglected.

(ii) Data derived from records of seismic bodily and surface waves, and free Earth oscillations. The data (i) and (ii) occupy a dominant, though not exclusive, place in the model constructions.

(iii) Evidence from a wide range of other sources, including data on Earth tides, thermal data, investigations on the variation of k with p, finite-strain and solid-state theory, laboratory experiments on rocks, including shock-wave experiments at pressures up to 4×10^{11} N/m^2, and evidence from geodesy, planetary physics, geology and geochemistry. This third body of evidence, though mostly less precisely determined than the seismic data, assists in assessing the plausibilities of models which fit the seismic data within the uncertainties, and usefully supplements the seismic data where the uncertainties are unusually large.

In the historical evolution of Earth models, density has been the key property. The distributions of other properties are fairly readily derivable when the density distribution has been determined. Attention will therefore first be devoted to the density determination.

2. THEORY ON DENSITY VARIATION

The density ρ in the Earth is a function of p, the temperature τ and parameters q_i representing chemical composition and phase. Thus

$$\frac{d\rho}{dz} = \frac{\partial\rho}{\partial p}\frac{dp}{dz} + \frac{\partial\rho}{\partial\tau}\frac{d\tau}{dz} + \Sigma\,\frac{\partial\rho}{\partial q_i}\frac{dq_i}{dz}, \qquad (2.1)$$

= P + T + Q, say. It transpires that the term P can be evaluated more accurately than T and Q. Also, T/P and Q/P are fairly small for most z. Hence the usual procedure has been to start by assuming $d\rho/dz = P$ and then proceed by successive approximation.

2.1 The Williamson-Adams equation

Let G be the gravitation constant and m the mass within the sphere of radius r and centre O. Since $\partial\rho/\partial p = \rho/k$, $dp/dz = g\rho = Gm\rho/r^2$, $\alpha^2 = (k + 4\mu/3)/\rho$ and $\beta^2 = \mu/\rho$, the equation $d\rho/dz = P$ becomes

$$d\rho/dz = Gm\rho/r^2\phi, \qquad (2.2)$$

where

$$\phi = \alpha^2 - 4\beta^2/3 = k/\rho. \tag{2.3}$$

An equation equivalent to (2.2) was used in theoretical work last century. It is now called the Williamson-Adams equation; these authors [2] substituted values of ϕ derived from seismic data into (2.2) to estimate $d\rho/dz$ numerically inside the Earth.

The application of (2.2) to an internal region of the Earth requires, in addition to knowledge of α and β, knowledge from non-seismic sources of values of ρ and m at some level of the region. For example, in treating the immediate subcrust, a variety of evidence has led to the assumption of about 3.3 g/cm^3 for the value ρ' of ρ at the top. Values of m as a function of r are derived starting from the surface, where $m = M$, and using $dm = 4\pi r^2\rho\, dr$ along with (2.2); the condition $m = 0$ at $r = 0$ has also to be satisfied.

2.2 Temperature correction

Birch [3] derived

$$T = -\gamma\vartheta\rho, \tag{2.4}$$

where γ is the coefficient of thermal expansion at constant pressure, and ϑ is the 'super-adiabatic' temperature gradient. (For a short derivation of (2.4), see Bullen [4].) For numerical details on the application of (2.4), see Bullen [5].

2.3 Generalization of the Williamson-Adams equation

Information on variations of chemical composition and phase in the Earth is not sufficiently well determined to enable the last term Q of (2.1) to be evaluated directly. But the following generalization (Bullen [6]) of the Williamson-Adams equation takes account of Q. It is to be understood below that $d\rho/dp$ stands for $(d\rho/dz)/(dp/dz)$; and similarly with dk/dp.

From (2.3), we have

$$(dk/dp)\,dp/dz = \phi\, d\rho/dz + \rho\, d\phi/dz,$$

whence, on putting $dp/dz = g\rho$ and dividing by ϕ,

$$d\rho/dz = \eta g\rho/\phi = \eta G m\rho/r^2\phi, \tag{2.5}$$

where

$$\eta = dk/dp - g^{-1}d\phi/dz. \tag{2.6}$$

The coefficient η also satisfies $d\rho/dp = \eta\rho/k$. The W.A. equation (2.2) is the particular case of (2.5) for which $\eta = 1$; in this case (Bullen [7])

$$dk/dp = 1 + g^{-1}d\phi/dz. \tag{2.7}$$

It is readily deduced (Bullen [8]) that

$$\eta = 1 - \frac{\gamma\vartheta\phi}{g} + \frac{\phi}{g\rho}\Sigma\frac{\partial\rho}{\partial q_i}\frac{dq_i}{dz}. \tag{2.8}$$

Thus positive values of ϑ reduce η, while continuous variation of chemical composition and/or phase in general raises η.

The utility of the generalization (2.5)-(2.6) lies in the fact that numerical information is available on all three of the items dk/dp, $d\phi/dz$ and g. Theoretical and laboratory evidence yields values of dk/dp within fairly close bounds at pressures in the Earth below about 1000 km depth; $d\phi/dz$ is determined (with uncertainties varying with z) from seismic data; and values of g are known for most z within fairly close bounds from early experience in constructing Earth models. Thus η can be estimated within assessable bounds in parts of the Earth where significant inhomogeneity is suspected. The stage has been reached where the inclusion of values of η in terms of z is a desirable part of the specification of Earth models.

3. THEORY ON COMPRESSIBILITY

Applications of theory of the type in §2, using appropriate observational data, led to an Earth model (Model A - Bullen [9]) which gave the distribution of ρ accurate within 5 per cent for $0 < z \le 5000$ km. The data included seismic bodily wave data on α and β, data derived from Earth tides on the fluidity of the outer core, along with auxiliary assumptions such as taking $\rho' \approx 3.3$ g/cm^3. Since α and β together give k/ρ and μ/ρ, the distributions of k and μ were immediately derivable to about the same accuracy. The distributions of p and g were directly derivable from the ρ distribution.

A striking feature of Model A was that, whereas ρ jumped by nearly 70 per cent and μ fell from a sizable value to virtually zero at the mantle-core boundary N, k and dk/dp were continuous at N within the uncertainties. This result led to the formulation of the 'k-p hypothesis' (Bullen [10]) to the effect that, to a useful first approximation, k and dk/dp are smoothly varying functions of p throughout the Earth below $z = 1000$ km, i.e. for $0.4 < p < 4 \times 10^{11}$ N/m^2, approximately. In respect of the variation of dk/dp, the hypothesis later received strong support from other sources, including finite-strain theory (Birch [3]). Later evidence has also placed fairly close bounds to the extent to which k is likely to deviate from smooth variation with p.

The k-p hypothesis places additional restrictions on the allowable variations of k, ρ and μ in various parts of the Earth. It entails solidity in the inner core (Bullen [10]) corresponding to the sizable jump in α found by Lehmann [11] from the outer to the inner core; this follows from the relation $\alpha^2\rho = k + 4\mu/3$ and the fact that to a first approximation ρ must increase with z throughout the Earth. The hypothesis also throws light on a likely abnormal density variation inside the lowest 200 km of the mantle.

In conjunction with Birch's estimate [12], from shock-wave experiments at pressures exceeding 10^{11} N/m^2, that the Earth's central density does not much exceed 13 g/cm^3, the k-p hypothesis enabled Earth model distributions to be reliably continued from z = 5000 km to the centre.

4. APPLICATION OF SEISMIC SURFACE WAVE AND FREE EARTH OSCILLATION DATA

The seismic bodily-wave data yield evidence on ρ, k and μ only in the combinations k/ρ and μ/ρ. The detail in §§2,3 enables evidence from outside seismology to be brought to bear in deriving values of ρ separately from k and μ. Seismic surface wave and free Earth oscillation data, however, provide independent evidence on ρ in combinations other than k/ρ and μ/ρ. This evidence has enabled some refinements to be added to Earth models constructed using the principles of §§2,3.

Examples of recent Earth models incorporating evidence from free Earth oscillations are the models HB_1 (Haddon and Bullen [13]) and B497 (Dziewonski and Gilbert [14]). The model HB_1 meets most requirements for the Earth's mantle, but has a simple fluid core. The model B497 used later observations of certain free Earth oscillation overtones to estimate rigidity in the inner core.

5. USE OF SEISMOLOGICAL INVERSION PROCEDURES

Until recently, the main approach to the construction of Earth models has been through successive approximation. For example, the model HB_1 was arrived at through a well-defined sequence, starting from the original Model A and incorporating successively evidence on the variation of k with p, evidence from shock-wave data to improve the lower-core density distribution, revised data on I, and data on free Earth oscillations. Successive approximation brought the reliabilities of models to the point where procedures towards further inprovements are often reducible to linear theory.

Some recent procedures aim to apply the total seismic data (bodily wave, surface wave and free Earth oscillation data) - i.e. to 'invert' the data - to arrive, independently of past Earth models, at ranges of values within which ρ, k and μ must lie for each z.

In one of the newer approaches (Keilis-Borok and others [15], Press [16]), huge numbers of models are randomly generated inside a computer by a Monte Carlo technique and subjected to tests which include fitting the different sections of seismic data within prescribed limits, fitting data on I, etc., and meeting various other stipulations, e.g. on the core-radius. Of the millions of models that may be generated, the computer prints out only a very small number which pass all the tests. Although freedom from bias is sometimes claimed for this procedure because of the absence of dependence on earlier Earth models, the procedure as so far applied has its own biases. For example, it has favoured models more complicated than the data warrant. In practice, it has, moreover, sometimes failed to find classes of models which are otherwise known to fit the data assumed. But the procedure is being developed and has considerable potentiality.

Other, more general, inversion procedures use sophisticated mathematics with the aim of covering comprehensively all models which are compatible with wide sets of data. The procedures sometimes introduce 'credibility' criteria with a view to arriving at an optimum model. Although considerable progress has been made with the auxiliary theory, the stage has not yet been reached where much information has been added to that derived through successive approximation. But some useful information has already been provided on the relative uncertainties of values of ρ, etc., at different depths. For details on various analytical aspects of general inversion procedures, see refs. 17-21.

6. CRITERIA FOR A STANDARD EARTH MODEL

The developments to date have made it desirable to formulate a standard Earth model for general reference purposes both inside and outside geophysics. A committee for this purpose was set up by the International Union of Geodesy and Geophysics in 1971.

A cardinal requirement of a standard model is simplicity. This requirement is so paramount that the specification of a standard model may well involve fewer parameters than the minimum number strictly demanded by the data. The problem of determining a standard model is therefore distinct from that of determining an optimum model on a given set of data.

Examples of points on which decisions have to be made are:

(i) What is the appropriate representation in a range of depth where there is evidence of a rapid or sudden change of property? Should the model show a mathematical discontinuity or a rapid continuous change? It is usual to treat the Mohorovičić discontinuity and the mantle-core and inner-core boundaries as mathematical discontinuities. But the decision is more difficult to make in other parts of the Earth's interior. 'When in doubt, smooth' has often been stated as a guiding principle, but the difficult question is really 'how smooth?'.

(ii) How should relative degrees of smoothing be determined among different properties? For example, should priority be given to simplicity in the variation of ρ or of β over a range of depth in which there is interplay ('trade-off') between ρ and β when a model is perturbed? Should η (§2.3) be kept constant inside a particular region of a model, or should η be let fluctuate as a consequence of smoothing procedures applied to α and β?

(iii) Should the possibility of a transition layer between the outer and inner core be ignored in a standard model?

(iv) Should β be taken constant in the inner core? It is to be noted that if β is constant, μ cannot be constant; for $\beta^2 = \mu/\rho$ and ρ is not constant.

For some further details on the problem of a standard Earth model, see ref. 22.

7. SOME NUMERICAL RESULTS

Table 1 gives values, derived from a selection of recent Earth models, of ρ, p, g, k, μ, α and β at various depths z (km) below the Earth's surface. The units are g/cm^3 for ρ; 10^{11} N/m for p, k and μ; m/s^2 for g; and km/s for α and β.

For further numerical detail, and a comprehensive account of the whole subject of Earth models, see Bullen [23].

Table 1. Model values of properties of the Earth's interior at selected depths z.

z	ρ	p	g	k	μ	α	β
0	2.84	0	9.82	0.65	0.36	6.3	3.6
30	3.32	0.01	9.84	1.07	0.72	7.80	4.65
200	3.39	0.06	9.90	1.39	0.69	8.26	4.50
400	3.70	0.14	9.96	1.89	0.84	8.92	4.72
650	4.17	0.23	9.99	2.68	1.43	10.48	5.80
1000	4.54	0.39	9.96	3.49	1.84	11.44	6.36
2000	5.09	0.87	10.02	5.07	2.44	12.79	6.92
2886	5.69	1.35	10.8	6.54	3.04	13.64	7.30
2890	9.95	1.35	10.8	6.54	0.00	8.12	0.00
4000	11.39	2.48	7.9	10.34	0.00	9.53	0.00
5120	12.70	3.34	4.4	13.50	0.00	10.33	0.00
5160	12.7	3.34	4.3	13.6	1.7	11.25	3.7
6371	13.0	3.67	0	15.0	1.3	11.25	3.2

REFERENCES

1. K. E. Bullen and R. A. W. Haddon, Phys. Earth Planet. Interiors 7, 199, 1973.
2. E. D. Williamson and L. H. Adams, J. Wash. Acad. Sci. 13, 413, 1923.
3. F. Birch, J. Geophys. Res. 57, 227, 1952.
4. K. E. Bullen, Trans. Amer. Geophys. Un. 34, 107, 1953.
5. K. E. Bullen, Mon. Not. R. Astr. Soc., Geophys. Suppl., 7, 214, 1956.
6. K. E. Bullen, Geophys. J., R. Astr. Soc., 7, 584, 1963.
7. K. E. Bullen, Mon. Not. R. Astr. Soc., Geophys. Suppl., 5, 355, 1949.
8. K. E. Bullen, Geophys. J., R. Astr. Soc., 13, 459, 1967.
9. K. E. Bullen, Introduction to the Theory of Seismology, University Press, Cambridge, 3rd ed, 1965.
10. K. E. Bullen, Nature, Lond., 157, 405, 1946.
11. I. Lehmann, Publ. Bur. Centr. Seismol. Internat. A, 14, 3, 1936.
12. F. Birch, Geophys. J., R. Astr. Soc., 4, 295, 1961.
13. R. A. W. Haddon and K. E. Bullen, Phys. Earth Planet. Interiors 2, 35, 1969.
14. A. M. Dziewonski and F. Gilbert, Geophys. J., R. Astr. Soc., 35, 401, 1973.
15. I. J. Asbel, V. I. Keilis-Borok and T. B. Yanovskaya, Geophys. J., R. Astr. Soc., 11, 25, 1966.
16. F. Press, J. Geophys. Res. 73, 5223, 1968.
17. L. B. Slichter, Proc. R. Soc. Lond. A, 224, 43, 1954.
18. G. Backus and F. Gilbert, Geophys. J., R. Astr. Soc., 16, 169, 1968.

19. H. Takeuchi and K. Sudo, J. Geophys. Res. 73, 3801, 1968.
20. V. I. Keilis-Borok (Ed), Computational Methods in Seismology (English trans. by E. A. Flinn, Consultants Bureau, New York), 1972.
21. L. Knopoff and D. D. Jackson, The analysis of undetermined and overdetermined systems, in course of publication.
22. R. D. Adams, K. E. Bullen, J. R. Cleary, A. M. Dziewonski, E. R. Engdahl, R. A. W. Haddon, A. L. Hales, R. Lapwood and others: A set of papers on Standard Earth Model, in course of publication, Phys. Earth Planet. Interiors.
23. K. E. Bullen, The Earth's Density, Chapman & Hall, London, to be published, April 1975.

THE Fe_2O THEORY OF PLANETARY CORES

K. E. Bullen

c/o Department of Applied Mathematics,
University of Sydney, Australia

ABSTRACT. A résumé is given of the evidence for the theory that the outer cores (when they exist) of terrestrial-type planets consist of the iron oxide Fe_2O, which is known to be unstable at ordinary pressures but stable at pressures equal to those in the Earth's core. The theory, while avoiding the main objections to the earlier phase-transition theory of planetary cores, permits the Earth, Venus and Mars to have a common overall composition. Essential to the theory is the assumption that the pressure at the Earth's mantle-core boundary is a critical pressure common to all planets which have outer cores. Brief comments are made on Mercury and the Moon.

1. THEORIES ON THE COMPOSITIONS OF THE CORES OF THE TERRESTRIAL PLANETS

By 1906, it was well established that the Earth has a dense central core, and in 1936 that this core consists of a (so-called) outer core and an inner core. An early view, principally based on meteorite evidence, was that the central core is composed of iron and nickel. Later investigations confirmed that this composition applies to the inner core with high probability, but indicated that the outer core has a density too low (at the pressures involved) to consist of pure iron and nickel. It was thereupon suggested that the outer core consists of iron alloyed with less dense elements (e.g. silicon, carbon, sulphur). On the hypothesis that the core consists predominantly of iron, and thus has a distinct chemical com-

Thoft-Christensen (ed.), Continuum Mechanics Aspects of Geodynamics and Rock Fracture Mechanics, 23-28.

position from the mantle, it was shown [1, 2] that the terrestrial-type planets Earth, Venus and Mars cannot have the same overall composition: for example, if Venus and Mars are assumed to possess similarly composed mantles and cores to those in the Earth, their mantle-core mass ratios would be 3.6 and 5.4, as against 2.1 for the Earth; these ratios would be entailed using observational data on the masses M and radii R of Venus and Mars. In the following, the subscripts V and M will indicate properties of Venus and Mars, respectively.

In 1948-9, it was shown by Ramsey [3] and Bullen [4] independently that if, in contrast to the predominantly-iron-core theory, the change at the Earth's mantle-core boundary N were a pressure phenomenon (the outer core thus consisting of a high-density metallic phase of the lower-mantle material), the observational values (at the time) of M_V, M_M, R_V and R_M, as well as the moment of inertia coefficient y_M, would be compatible with Earth, Venus and Mars having the same overall composition. An essential point is that if the phase transition occurs at the same critical pressure p_c in all three planets, the mantle-core mass ratio would increase with decreasing planetary size.

The phase-transition theory appeared at the time to fit all the relevant observational data remarkably well, but the theory later met several difficulties, chiefly: (i) having regard to the packing of oxygen atoms in the Earth's lower-mantle material (presumed to have a composition equivalent to that of ultra-basic rock), many geochemists found difficulty in reconciling the theory with the large density jump (in a ratio about 0.7) at N; (ii) transitory shock-wave experiments at core pressures (though not definitive on the point) have failed to supply positive evidence that the phase transition does occur; (iii) the observational estimate of R_V has been substantially reduced from 6200 $\pm$ 50 km around 1950 to 6050 $\pm$ 10 km in 1972. (The phase-transition theory had predicted $R_V = 6270$ km.)

The Fe_2O theory described below avoids the main objections to the phase-transition theory, while preserving the notion that the pressure p_c at N is a critical pressure relevant to all terrestrial-type planets. The Fe_2O theory restores compatibility with the planetary observational data and the notion of a common overall composition of Earth, Venus and Mars.

For further details on events leading to the formulation of the Fe_2O theory, see ref. 5.

2. THE Fe_2O THEORY

Soroktin [6] inferred from physical chemistry calculations that the iron oxide Fe_2O, which is unstable at ordinary pressures, becomes stable at a pressure of the order of that at the Earth's mantle-core boundary. His calculations gave a pressure-density relation for Fe_2O matching that in the Earth's outer core, leading him to propose an Fe_2O composition for this region. He associated the occurence of Fe_2O with a process essentially equivalent to

$$2FeO \rightarrow Fe_2O + O, \tag{2.1}$$

and applied the idea to constructing an Earth model in which the oxygen thus released plays a major rôle.

The Fe_2O theory of the writer [7] follows Soroktin in postulating that the Earth's outer core is composed of Fe_2O, but then deviates from Soroktin by associating the occurrence of Fe_2O with the relation

$$Fe_2O \Leftrightarrow FeO + Fe. \tag{2.2}$$

The reason for proposing (2.2) in place of (2.1) is that (2.2) leads to interesting implications on the cores of other terrestrial-type planets.

Stripped to essentials, the Fe_2O theory envisages a theoretical family of terrestrial-type planets, including the Earth, all with the same overall chemical composition, which involves two primary materials: a basic mantle material, X say, and Fe_2O. The equivalent masses of X and Fe_2O are in the same ratio for all the planets, but the Fe_2O exists as such only inside those planets where the critical pressure p_c is reached. In planets where insufficient pressures are reached, some or all the Fe_2O that would otherwise occur is broken down into FeO and Fe in accordance with (2.2), the FeO going into the mantle to mix with X, and the Fe sinking to form an inner core. (The composition of X need not here be specified; X may already contain some FeO, apart from the additional FeO arising from the breakdown of the Fe_2O.)

In these circumstances, there would be three subsets of planets: K, J and H, say. The larger planets (K) would have mantles composed purely of X, 'outer cores' of Fe_2O, and no inner cores. The smaller planets (H) would have mantles composed of X and FeO mixed, cores ('inner cores') of Fe, and no outer cores. The intermediate planets (J) would have mantles composed of X and some FeO, outer cores of Fe_2O (not all of which has broken down), and inner cores of Fe. With K and J, which have outer cores, the pressure at the base of the mantle is p_c in all cases; with H, this pressure is always less than p_c. The mantle-core mass ratios are the same for all K planets, and for all H planets (but the ratio is different for K and H).

Earth and Venus would correspond to members of J, Mars of H. No known planet corresponds to K, so that observational tests of the theory have to be limited to J and H. It is incidentally to be expected that, as in the Earth, the outer cores of the J planets would be fluid, and the inner cores of the J and H planets solid.

3. PROCEDURE IN CALCULATIONS

Subject to certain auxiliary assumptions, based on relevant evidence but avoiding the less important complications, the Fe_2O theory can, for member planets of assigned M, be applied to calculating their density distributions uniquely and hence the values of R and y. The theory can then be tested by comparing with the observational M, R and y.

Examples of the auxiliary assumptions are: pressure-density equations of state for particular materials in the interior of the Earth apply to the same materials in the other planets; the masses of Fe in the inner cores and of additional FeO in the mantles of the J and H planets are always in the proportion 7/9, which is such that the Fe and additional FeO would combine precisely to form Fe_2O; the presence of nickel in cores is ignored (an auxiliary calculation showed the effect of this to be quite slight); possible chemical interaction between X and additional FeO in the mantle is ignored (this effect would also probably be slight in the present context). Even with these simplifications, the calculations are fairly complex, and so compressibilities in particular planetary zones were neglected in a first approximation. Some progress has been made with a second approximation in which compressibilities are taken into account.

4. COMPARISON WITH OBSERVATION

For Venus, assuming $M_V = 4.87 \times 10^{24}$ kg, the calculations gave $R_V = 6010$ km in the first approximation. The second approximation, with due allowance for uncertainties in the procedures for incorporating compressibilities, gave full agreement with the observational R_V (6050 km). There is no observational value available for y_V.

The thickness yielded for the (presumed fluid) outer core of Venus was a little less than 600 km, in contrast to 920 km on the phase-transition theory and 2000 km on the predominantly-iron-core theory. On current ideas on the Earth's magnetic field, the Fe_2O theory thus accounts best for the failure to observe a significant magnetic field around Venus.

For Mars, assuming $M_M = 0.642 \times 10^{24}$ kg, the results of the calculations agreed closely with the observed R_M (3388 ± 5 km) and y_M (0.376) on the first approximation. Allowance for compressibility somewhat reduced the fit, giving $R_M \approx 3365 \pm 10$ km and $y_M \approx 0.364$. At least a large part of the apparent discrepancies with observation could be attributed to appreciable deviatoric stresses inside Mars, the calculations having assumed hydrostatic conditions. In hydrostatic conditions, $y_M = 0.376$, 0.364 would correspond to polar flattenings of 0.0053, 0.0050, respectively.

The phase-transition theory would fit the observational M_M, R_M and y_M closely if Mars were in a nearly hydrostatic state, but gives a much inferior fit to the Fe_2O theory with M_V and R_V. Both theories (as well as the iron-core theory) are compatible with Mars being solid throughout and thus agree with the failure to detect a magnetic field around Mars.

The Fe_2O theory has the great advantage over the phase-transition theory that it avoids all the objections (i), (ii) and (iii) of § 1, as well as giving good agreement with the observational data.

5. FURTHER REMARKS

The Fe_2O theory does not fir the observational M and R for Mercury, as Mercury is now. Neither does any other theory that is not both ad hoc and complex. But there is strong likelihood that because of Mercury's high temperature in the vicinity of the sub-solar point, a sizable part of the mantle has volatilized away, leaving the present Mercury appreciably denser than the primitive Mercury, and so accounting reasonably for the abnormally high observed mean density of the present Mercury. A calculation based on the Fe_2O theory has incidentally shown that if the primitive Mercury had the characteristics of an H planet, its mass would have been about 1×10^{24} kg, which is appreciably greater than the mass of Mars.

The observational mass and radius of the Moon are incompatible with the Moon being a member of the family of planets envisaged above. But on nearly all theories, the Moon appears to have had an exceptional origin. Only the phase-transition theory allows the Moon to have the same overall composition as the Earth.

An important aspect of the Fe_2O theory is that it lends itself to possible future test when seismological data come to be gathered on Mars and (more distantly) Venus. Seismological evidence on the radius of a core in Mars and on the thickness of a fluid outer core in Venus could discriminate crucially between the Fe_2O and other theories.

For further detail on the Fe_2O theory, see refs. 5, 7, 8.

REFERENCES

1. H. Jeffreys, Mon. Not. R. Astr. Soc., Geophys. Suppl., **4**, 62, 1937.
2. K. E. Bullen, Rept. Austral. N. Z. Assoc. Adv. Sci. **23**, 25, 1937.
3. W. H. Ramsey, Mon. Not. R. Astr. Soc. **108**, 406, 1948.
4. K. E. Bullen, Mon. Not. R. Astr. Soc. **109**, 457, 1949.
5. K. E. Bullen, The Earth's Density, Chapman and Hall, London, to be published in April 1975.
6. O. G. Soroktin, personal communication, 1971.
7. K. E. Bullen, Nature, Lond., **243**, 68 and **246**, 85, 1973.
8. K. E. Bullen, The Moon **7**, 384, 1973.

PRINCIPLES OF FRACTURE MECHANICS

F. Erdogan

Lehigh University, Bethlehem, Pa. 18015

1. INTRODUCTION

When a solid is subjected to a certain system of external loads under a given set of environmental conditions, in most cases the response is simply one of deformation in the sense that topologically the medium remains unchanged. In this case, at least conceptually, the solution of the problem presents no difficulty, that is, after the appropriate material characterization it can be formulated as an initial-boundary value problem and all the desired field quantities may be obtained upon solving the related field equations. On the other hand, in addition to deformations taking place in the body, if new surfaces are created within the medium as a result of excessive applied loads, the body is said to undergo fracture. From the viewpoint of physical applications and mathematical modeling, in this case one generally distinguishes two types of problems, namely fracture initiation and fracture propagation. For a given solid with a given defect geometry, in the former one needs to evaluate the critical level of the applied loads corresponding to the inception of the fracture process. Since no new surfaces have as yet been created, here the related mechanics problem is again a deformation type of problem complemented by an appropriate fracture criterion which relates the inherent fracture resistance of the solid to the applied loads and geometry of the medium.

On the other hand in the problem of fracture propagation for the general case even the formulation of the problem does not appear to be feasible. However, if the material is homogeneous and if the geometry of the medium and the applied loads satisfy certain symmetry conditions in such a way that the fracture surface A(t) may be

Thoft-Christensen (ed.), Continuum Mechanics Aspects of Geodynamics and Rock Fracture Mechanics, 29-44.

characterized in terms of a single length parameter a(t), then an acceptable formulation of the problem would be as follows:

$$L_i(u_1,u_2,u_3,a) = \rho \frac{\partial^2 u_i}{\partial t^2}, \quad (i = 1,2,3),$$

$$L_o(u_1,u_2,u_3,a) = 0 , \qquad (1.a,b)$$

subject to appropriate initial and (time-dependent) boundary conditions. In (1), u_1, u_2, u_3 are the components of the displacement vector, ρ is the mass density, L_1, L_2, L_3 are the differential operators of the related deformation problem, and the operator L_o which accounts for the additional unknown function a(t) is the fracture criterion for the propagating fracture. The operator L_o also contains the "material constants" representing the fracture resistance of the medium. It is thus clear that in order to deal with the problem of fracturing solids, the concept of fracture, its mechanism, and its causes have to be incorporated quantitatively into the mechanics of materials through the development and introduction of a reliable and practical fracture criterion. This in turn requires an extensive study of the fracture phenomenon which has been the subject of a variety of fracture theories.

2. FRACTURE THEORIES

Since the fracture process is basically the formation of new surfaces in the solid, at the most basic level the essential feature of the phenomenon is the rupture of (cohesive or adhesive) interatomic bonds in the material. At the other end of the scale the phenomenon is exhibited in the form of rupture separation of a full scale structural member as a result of the propagation of a dominant flaw in the medium. In between the fracture process may be considered as the nucleation and growth and/or coalescence of micro- and macro-voids or cracks. Thus in studying the fracture of solids ideally one may have to consider such widely diverse factors as the microscopic phenomena taking place at locations of fracture nucleation, the microstructure of the material, its nature and its composition, and the macroscopic effects regarding the gross loading and environmental conditions and the geometry of the medium. Due to this highly complex nature of the phenomenon, at the present time there appears to be no consistent single theory dealing satisfactorily with all its relevant aspects. Thus, partly due to the highly diverse nature of the phenomena leading to the fracture process and partly to the diversity in the background and interest of the investigators studying the subject, generally speaking the existing theories treat the subject from only one of three points of view, namely, microscopic (or atomic), microstructural, and continuum or macroscopic.

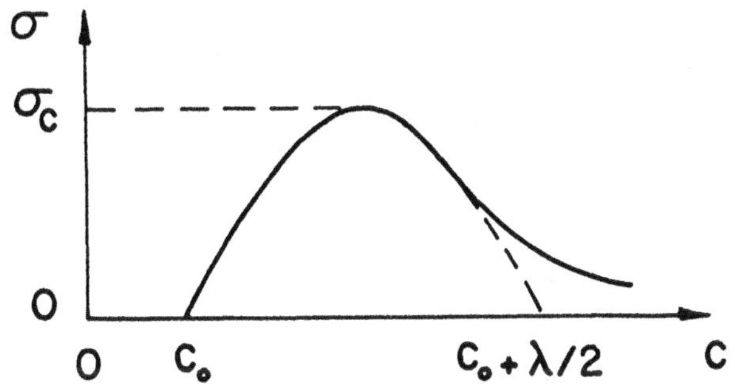

Fig. 1. Interatomic force vs. atomic distance.

2.1 Microscopic aspects of fracture

Here the main interest is in the rupture of interatomic bonds, the tools being that of solid state and quantum physics and chemistry. The process is far from being fully understood. However, to fix the ideas it is instructive to consider the following highly simplified model: Assuming the necessary symmetry in lattice structure and fracture front and ignoring all the nonlocal effects, let the interatomic force shown in Figure 1 be approximated by the following half-sine relation:

$$\sigma \simeq \sigma_c \sin \frac{2\pi(c-c_o)}{\lambda} , \tag{2}$$

where c is the interatomic distance, c_o is the value of c corresponding to zero applied load* σ, σ_c is the theoretical cohesive strength of the material, and λ is a material constant. Defining the Young's modulus as

$$E = \lim_{c \to c_o} \frac{d\sigma}{d\left(\frac{c-c_o}{c_o}\right)} = \frac{2\pi}{\lambda} \sigma_c c_o , \tag{3}$$

the theoretical strength becomes

$$\sigma_c = E/(2\pi c_o/\lambda) . \tag{4}$$

On the other hand the energy released as a result of the rupture of atomic bonds (per unit area) may be expressed as

---

* Here σ is the interatomic force when referred to a single pair and may be considered as "stress" when referred to a "unit" area of ruptured bonds.

$$\frac{dU}{dA} = \int_{c_o}^{c_o+\lambda/2} \sigma_c \sin \frac{2\pi}{\lambda} (c-c_o)dc = \frac{\lambda\sigma_c}{\pi} . \tag{5}$$

If γ_s is the specific surface energy of the material, observing that the bond rupture creates two surfaces, from (5) it follows that

$$2\gamma_s = \frac{\lambda\sigma_c}{\pi} . \tag{6}$$

Since γ_s, E, and c_o are more easily measurable quantities, from (4) and (6) eliminating λ we find

$$\sigma_c = (E\gamma_s/c_o)^{1/2} . \tag{7}$$

In real solids more refined calculations indicate that $\sigma_c \simeq (1/4 - 1/13)E$. Of course the actual strength σ_a of the solid is much lower than σ_c. This is generally attributed to microstructural defects in the solid. Thus, if one assumes that 2a is the characteristic length of the microstructural defects in the solid (assumed to be an elliptic hole which is atomic sharp, i.e., $\rho \simeq c_o$, ρ being the radius of curvature), one may obtain a rough estimate for σ_a by simply using the elasticity solution, which gives

$$\sigma_{max} = \sigma_a(1 + 2\sqrt{a/\rho}), \text{ or } \quad \sigma_a \simeq \frac{1}{2}\sigma_c(c_o/a)^{1/2} , \tag{8}$$

which is of the correct order of magnitude.

There are other microscopic fracture theories mostly based on the approach of statistical mechanics. Even though the microscopic theories provide some phenomenological insight into the explanation of the fracture process, they fail to provide a satisfactory quantitative working tool for the physical applications.

2.2 Microstructural fracture theories

These are the theories dealing with the phenomena which take place in the solid within approximately $10^{-6} - 10^{-3}$ cm. length scale (motion of dislocations, formation of subgrain precipitates, subgrain slip, formation of slip bands, formation of intrusions and extrusions, formation of microcracks, and microcrack propagation in crystalline materials, and formation of voids through slip and bond rupture in the chains and between the chains of molecules in the amorphous materials). These theories too are largely qualitative and aim at explaining the mechanism of the crack formation rather than trying to develop quantitative fracture criteria.

2.3 Continuum theories of fracture

The objective of the macroscopic theories of fracture which are based on the notions of continuum mechanics and classical thermodynamics is to provide the quantitative working tools applicable to the fracture of structural materials. In the macroscopic approach to fracture it is implicitly assumed that the material contains some macroscopic flaws which may act as fracture nuclei and the medium is a homogeneous continuum in the sense that the size of the macroscopic flaws is large compared with the characteristic microstructural dimension of the material. The problem is then to study the effect of the applied loads, the flaw geometry, and the environmental conditions on the fracture of the solid - a subject which has come to be known in the past two decades as fracture mechanics. In fracture mechanics there are generally three types of problems which, even though related, may be studied more or less independently:

(a) The experimental work aimed at determining the inherent resistance of the material to fracture (the quantity known as the "characteristic strength parameter") corresponding to the conjectured mode of fracture, such as "brittle" or "quasi-brittle" fracture, "ductile" fracture, and debonding or delamination.

(b) The solution of the mechanics problem for the given geometry and applied loads in order to determine the "critical load factor" representing the severity of the applied loads and corresponding to the conjectured mode of fracture.

(c) The development of an appropriate fracture theory and a related quantitative "fracture criterion". This is usually nothing but the direct comparison of the "characteristic strength parameter" of the material with the calculated "critical load factor". It is then clear that in fracture studies the selection of a proper fracture criterion must precede the experimental determination of the strength parameter and the theoretical evaluation of the load factor. The remainder of this paper will be devoted to a brief review of some of the current macroscopic fracture theories.

3. THE ENERGY BALANCE THEORY

Consider a solid subjected to a given system of external loads and containing a "dominant flaw" from which the fracture is most likely to initiate and which may be considered as a plane crack. In the most general case, the thermodynamic equilibrium of the body requires that

$$\frac{dU}{dt} = \frac{dV}{dt} + \frac{dT}{dt} + \frac{dD}{dt} , \qquad (9)$$

where t is the time, U is the work done by the external loads, V is the recoverable (elastic) component of the stored energy, T is the kinetic energy, and D is the sum of all the irreversible energies such as the surface tension, plastic work, viscous dissipation, etc. If we assume that S(t) is the total surface, A(t) is the crack surface, S_o is the surface of the solid excluding the crack which (for the purpose of fracture analysis) remains constant as the crack propagates, and all the energy dissipation takes place around the advancing periphery of the crack, we may write

$$S = S_o + A(t) , \quad \frac{dD}{dt} = \frac{dD}{dS}\frac{dS}{dt} = \frac{dD}{dA}\frac{dA}{dt} = \gamma_F \frac{dA}{dt} . \qquad (10)$$

The meaning of $(dD/dA) = \gamma_F$ is the energy required to create a unit amount of fracture surface, and is known as the fracture energy of the solid. For structural materials γ_F is considerably higher than the surface tension energy of the solid ($\gamma_F \sim 10^3 \gamma_s$).

In the quasi-static case, noting that $dT/dt = 0$, $d(\ )/dt = (d(\ )/dA)dA/dt$, (9) becomes

$$\frac{d}{dA}(U-V) = \gamma_F . \qquad (11)$$

In (11) the left-hand side represents the internally released or externally added energy (i.e., the energy available for fracture) per unit area of new fracture surface. Thus, (11) may be considered a very simple and also a very general fracture criterion. From the physical meaning of the terms in (11) it is clear that for constant γ_F the stability of fracture propagation may be determined from

$$\frac{d}{dA}\left[\frac{d}{dA}(U-V) - \gamma_F\right] \begin{cases} > 0 : & \text{unstable fracture,} \\ < 0 : & \text{stable fracture,} \\ = 0 : & \text{neutral equilibrium.} \end{cases} \qquad (12)$$

Equation (11) basically states the fracture criterion as proposed by Griffith in his 1920 paper [1], except that in the ideal brittle materials he considered, γ_F was the surface tension energy γ_s. Since the practical fracture resistance of the structural solids was observed to be orders of magnitude higher than the corresponding γ_s, Griffith's work was more or less ignored until the late 1940's. The revival of the theory came about after the X-ray work indicated that even in materials which were fracturing in "purely brittle" fashion there were extensive plastic deformations on the fracture surfaces [2] which led Irwin to propose that in the energy balance in addition to the surface tension energy, the plastic work should also be considered as a dissipative energy component [3]. The same argument appears to have been stated independently by Orowan some years later [4].

As some simple examples for the application of the energy balance criterion consider the following problems:

(a) Splitting of beams or plates:

The problem is described by Figure 2. If the load P is constant, as $a \to a + da$, $\ell \to \ell + d\ell$, and it may easily be shown that

$$dU - dV = \frac{1}{2} P d\ell \ , \quad \ell = 2 \frac{Pa^3}{3E_1 I} \ ,$$

$$dU - dV = \frac{P^2 a^2}{E_1 I} da \ , \quad \gamma_F dA = \gamma_F b \, da \ ,$$

$$\frac{P_c^2 a^2}{bE_1 I} = \gamma_F \ , \tag{13}$$

$$\frac{d}{dA}\left[\frac{d}{dA}(U-V) - \gamma_F\right] = \frac{d^2}{b^2 da^2}(U-V) = \frac{2P^2 a}{b^2 E_1 I} > 0 \ , \tag{14}$$

beam: P = total load, $E_1 = E$, $I = bh^3/12$,

plate: P = load/unit width, $E_1 = E/(1-\nu^2)$, $I = h^3/12$, $b = 1$.

That is, the fracture process is unstable and for a given crack length a the fracture load P_c is determined by (13).

On the other hand if the displacement ℓ is constant (i.e., the case of wedging), as $a \to a + da$, $P \to P + dP$ ($dP < 0$), and we have

$$dU = 0 \ , \quad dV = \frac{1}{2}\ell dP \ , \quad P = \frac{3E_1 I\ell}{2a^3} \ , \quad dP = -\frac{9E_1 I\ell}{2a^4} da \ ,$$

$$dU - dV = \frac{9E_1 I\ell^2}{4a^4} da \ , \quad dA = b\,da \ ,$$

$$\gamma_F = \frac{9E_1 I\ell_c^2}{4ba^4} \quad \left(= \frac{P_c^2 a^2}{bE_1 I}\right). \tag{15}$$

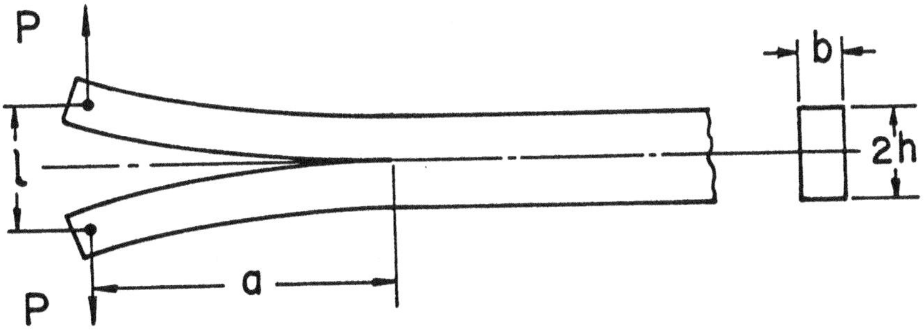

Fig. 2. Splitting of a beam.

$$\frac{d}{dA}\left[\frac{d}{dA}(U-V) - \gamma_F\right] = -\frac{9E_1 I \ell^2}{b^2 a^3} < 0 . \quad (16)$$

That is, the fracture process is stable and (15) gives the corresponding equilibrium values of ℓ_c and a at fracture.

(b) Tearing of a long strip:

In the example shown in Figure 3, the external load is applied through rigid clamps which are held fixed at $\varepsilon_{yy} = \varepsilon_o$ when the material begins to fracture. The fracture process is "self-similar" in the sense that the stress state within the perturbation zone around the crack tip remains unchanged as the crack propagates. Thus, one may easily distinguish three different zones, namely, the unloaded zone in which the strain energy density W is zero, the perturbation zone in which W is unknown, and the loaded zone in which W is constant. As $a \to a + da$ we then have

$$dU = 0 , \quad dV = - W d\tau = - \frac{1}{2} \frac{E\varepsilon_o^2}{1-\nu^2} 2bh\, da ,$$

$$dA = 2h\, da , \quad dU - dV = 2\gamma_F h\, da ,$$

$$2\gamma_F = \frac{bE\varepsilon_o^2}{1-\nu^2} , \quad (17)$$

$$\frac{d}{dA}\left[\frac{d}{dA}(U-V) - \gamma_F\right] = 0 , \quad (18)$$

indicating that the fracture process is in neutral equilibrium and the critical strain ε_o maintaining the fracture is given by (17).

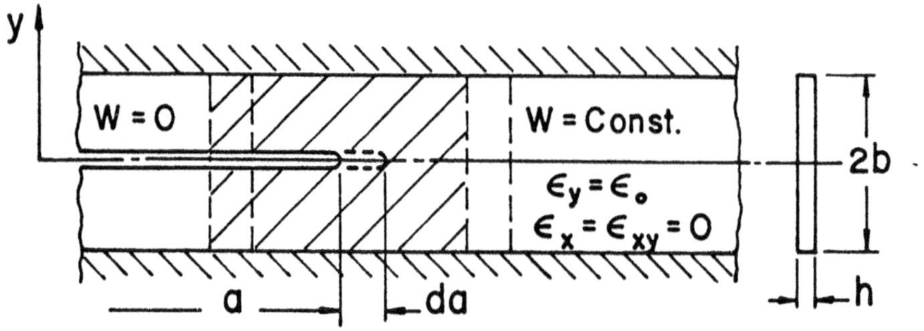

Fig. 3. Tearing of an infinite strip.

4. THE STRESS INTENSITY FACTOR

Consider now a linearly elastic solid containing a dominant flaw which may be represented by a plane crack having a smooth curve as a boundary. Upon loading, some inelastic deformations will develop around the edge of the crack. This toroidal region along the crack front in which all the energy dissipation is assumed to take place will be called the "fracture process zone". Let the characteristic size of the crack and the process zone size be a and p, respectively. If $p \ll a$, then it is reasonable to assume that the energy (U-V) "pumped" into the fracture zone will come from the elastic bulk of the solid and therefore will not be critically dependent on the stress state in the close neighborhood of the crack periphery. The significance of this observation lies in the fact that one may now be justified in calculating the energy (U-V) available for fracture from an ideally elastic medium.

Figure 4 shows a section of the medium perpendicular to the crack front before and after loading. For a linearly elastic solid after loading the crack profile near the front becomes parabolic. If the external loads and the overall geometry have a symmetry with respect to the plane of the crack, from the elasticity solution of the problem for $r = a-x \ll a$ and $\rho = x-a \ll a$, one may write the following asymptotic relations for the crack surface displacement and the cleavage stress (Figure 4b):

$$v(x,0) = \frac{2}{E_1} k_1 \sqrt{2(a-x)} \, [1 + 0(\frac{a-x}{a})] , \quad (x < a), \tag{19}$$

$$\sigma_{yy}(x,0) = \frac{k_1}{\sqrt{2(x-a)}} [1 + 0(\frac{x-a}{a})] , \quad (x > a). \tag{20}$$

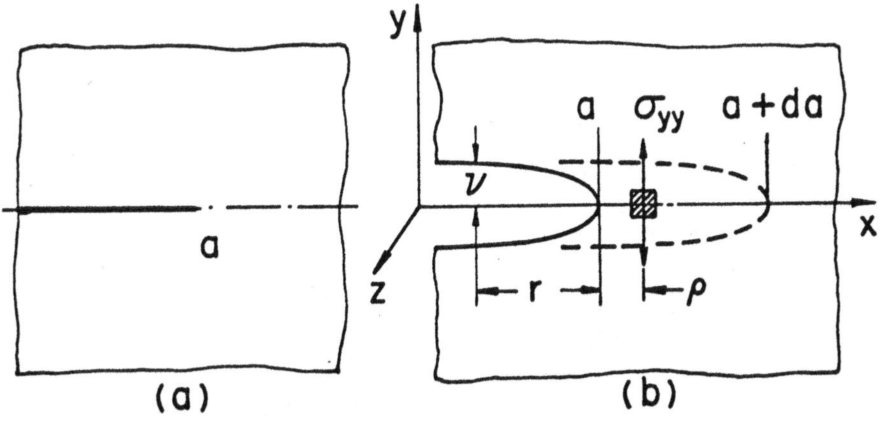

Fig. 4. Geometry of the propagating crack front.

$$E_1 = \begin{cases} E : \text{ plane stress,} \\ E/(1-\nu^2) : \text{ plane strain.} \end{cases}$$

The constant k_1 is a function of the geometric dimensions and the external loads and is linearly dependent on the load amplitude. Thus, k_1 is a measure of the severity of the loading and geometry and is known as the stress intensity factor (after Irwin).

Let now the crack front advance parallel to itself by an amount da (Figure 4b) (under fixed grip conditions). The externally added (dU) and internally released energy (dV) per unit crack front may then be expressed as follows:

$$dU = 0 ,$$

$$dV = -2 \int_a^{a+da} \frac{1}{2} \sigma_{yy}(x) v(x-da) dx$$

$$\simeq - \int_a^{a+da} \frac{k_1}{\sqrt{2(x-a)}} \frac{2k_1}{E_1} \sqrt{2(a+da-x)} \, dx = - \frac{\pi k_1^2}{E_1} da ,$$

Noting that for unit crack front dA = 2da, the fracture criterion becomes

$$\frac{d}{da}(U-V) = \frac{\pi k_1^2}{E_1} = 2\gamma_F . \tag{21}$$

Since U-V is a "potential" and a is a "distance", the derivative in (21) has the dimension of "force". Consequently

$$G = \frac{d}{da}(U-V) = \frac{\pi k_1^2}{E_1} \tag{22}$$

is also known as the "crack extension force" (after Irwin). In current fracture mechanics terminology $2\gamma_F$, the energy required per unit crack extension, is generally known as the fracture toughness, G_C (G_{IC} for plane strain, again after Irwin). Strictly speaking, (21) is applicable only if the fracture zone size p is small compared to the crack size a, indicating that the resulting fracture surface is generally flat and is associated with plane strain fracture (i.e., the crack propagates in its own plane rather than in a plane inclined 45° to the original plane of the crack; the latter fracture is known as shear fracture and is generally associated with plane stress configurations). In applications to flat fracture the fracture criterion (21), i.e.,

$$G \leq G_{IC} , \tag{23}$$

is also used in the form

$$k_1 \leq k_{IC} , \quad k_{IC} = \sqrt{\frac{2\gamma_F E_1}{\pi}} = \sqrt{\frac{G_{IC} E_1}{\pi}} , \tag{24}$$

where the material's new fracture resistance parameter k_{IC} is known as the "critical stress intensity factor".

In his investigations of the fracture strength of glass plates, the original problem Griffith considered was an (infinitely) large plate containing a straight through crack of length 2a (located at $y = 0$, $-a < x < a$) and subjected to a uniform stress $\sigma_{yy}^{\infty} = \sigma_o$ perpendicular to the plane of the crack [1]. The solution of the problem was given by Inglis [5], from which we may write

$$\sigma_{yy}(x,0) = \frac{\sigma_o |x|}{(x^2-a^2)^{1/2}} , \quad (|x| > a) . \tag{25}$$

From the definition of the stress intensity factor as given by (20) and from (25) it follows that

$$k_1 = \lim_{x \to a} \sqrt{2(x-a)} \, \sigma_{yy}(x,0) = \sigma_o \sqrt{a} . \tag{26}$$

At the onset of fracture $\sigma_o = \sigma_a$ and from (24) and (26) we have

$$\sigma_a = \sqrt{\frac{2\gamma_F E_1}{\pi a}} . \tag{27}$$

Replacing γ_F by known surface tension energy γ_s and measuring the rupture strength σ_a, Griffith indicated that (27) may be used to determine the characteristic defect size 2a in the material.*

For the problem described by Figure 3, from (17) and (21) (assuming $E_1 = E$) one easily obtains

$$k_1 = E\varepsilon_o \sqrt{\frac{b}{\pi(1-\nu^2)}} . \tag{28}$$

Needless to say, the elasticity solution of this problem is rather complicated; the simple result given by (28) is due to the self-similar nature of the crack propagation problem.

It should be noted that from (12) and (21) the fracture stability may be expressed in terms of k in the following simple form:

---

* If the defect is assumed to be an internal "penny-shaped" crack of radius a, then instead of (26) one should use $k_1 = \frac{2}{\pi} \sigma_o \sqrt{a}$, which comes from the solution of the axisymmetric problem [6].

$$\frac{\partial k_1}{\partial a} \begin{cases} > 0 : & \text{unstable fracture,} \\ < 0 : & \text{stable fracture,} \\ = 0 : & \text{neutral equilibrium,} \end{cases} \tag{29}$$

where it is assumed that $k_1 > 0$ (i.e., rupture may occur only under tension). Thus, (26) shows that the so-called Griffith crack is unstable.

If the medium lacks symmetry in loading and geometry with respect to the plane of the crack, $y = 0$ (Figure 4), then on the $y = 0$ plane the shear stresses σ_{xy} and σ_{yz} will not be zero and around the crack front for $(x-a) << a$ the elasticity solution gives the following asymptotic relations:

$$\sigma_{yy}(x,0) = \frac{k_1}{\sqrt{2(x-a)}}\,[1 + 0(\frac{x-a}{a})] \ ,$$

$$\sigma_{xy}(x,0) = \frac{k_2}{\sqrt{2(x-a)}}\,[1 + 0(\frac{x-a}{a})] \ ,$$

$$\sigma_{yz}(x,0) = \frac{k_3}{\sqrt{2(x-a)}}\,[1 + 0(\frac{x-a}{a})] \ , \quad (x > a). \tag{30.a-c}$$

Here k_1, k_2, and k_3 respectively are known as the opening (mode I), forward shear (mode II), and antiplane shear (mode III) components of the stress intensity factor.* In this case for a crack propagating in its own plane, the energy available for fracture (per unit crack front) replacing (21) becomes

$$\frac{d}{da}(U-V) = \frac{\pi}{E_1}\,(k_1^2 + k_2^2) + \frac{\pi}{\mu}\,k_3^2 \ , \tag{31}$$

where μ is the shear modulus of the material.

5. CRITICAL COD AND PLASTIC INSTABILITY

Even though the underlying principle of the energy balance criterion is quite general, quantitatively it gives good results only for "plane strain" type of configurations where the size of the fracture process zone surrounding the leading edge of the crack is small in comparison with the characteristic crack size. In practice the difficulty lies in the calculation of the load factor G (eq. 22)

---

* The stress intensity factors K_1, K_2, K_3 which appear in the literature are related to those defined by (30) through $K_i = k_i\sqrt{\pi}$, $(i = 1,2,3)$.

with sufficient accuracy and in properly isolating and measuring the strength parameter G_C. Even though developed only for non-linear elastic (hence, nondissipative) materials, recent attempts of computing G through path-independent integrals (known as the J-integral) have met with partial success [7 – 9]. However, the problem of a reliable strength characterization for G_C still remains very elusive. Hence, in problems involving the presence of large scale plastic deformations around the crack front, the energy balance criterion is of limited use.

The recent investigations on the development of the "crack extension resistance curve" for (or the so-called K_R-characterization of) the fracture of sheet materials appear to be promising [10]. However, in this case too the isolation of a geometry and size-independent characteristic strength parameter has not yet been successful.

Contending that fracture is basically rupture-separation of the material around the leading edge of the crack, particularly in the presence of large scale plastic deformations, instead of considering the global energy balance some investigators have concentrated their attention mostly on the phenomena taking place at the crack tip region. The main argument here is that in such cases perhaps a true measure of the severity of the applied loads and the geometry may be the deformation state at the crack front characterized by the amount of stretch, δ (or the "crack opening displacement" COD) suffered by the layer of material in the plane and ahead of the crack front. In turn the fracture resistance of the material may be characterized by the value of this stretch at rupture, δ_C (or critical COD), the proposed strength parameter, which is conjectured to be (with certain qualifications) a material constant [11 - 13]. In practice δ_C may be measured from standard experiments without too much difficulty. As for the evaluation of the load factor δ, the "plastic strip model" (known variously as Dugdale, -Barenblatt, -Muskhelishvili model) appears to give very satisfactory results. The model is a highly oversimplified representation of the deformation state around the crack front and is based on the assumption that the plastic deformations take place only in a thin layer of the material in the plane of the crack. Thus removing this layer, replacing it by tensile tractions of magnitude σ_Y (the yield strength), and assuming that the (fictitious) crack front is at the border of the plastic zone, the elasticity solution gives δ as the crack surface displacement at the (actual) crack tip.

For example, for a large plate under unidirectional tension shown in Figure 5, observing that at the border of the plastic zone $x = a+p$ the stress state must be finite, and hence k_1 must vanish, through superposition we obtain

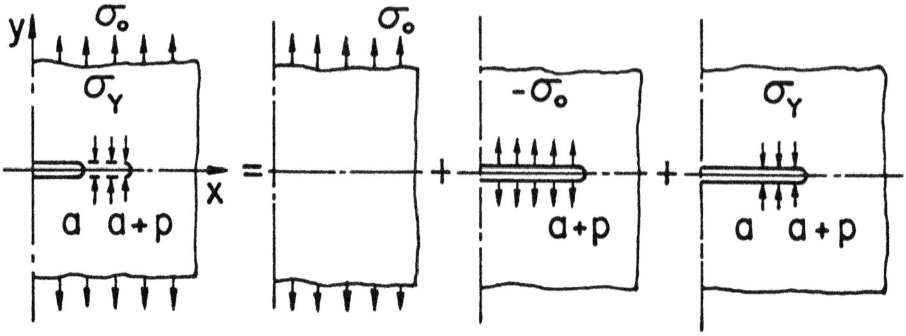

Fig. 5. Superposition for the calculation of COD.

$$k_1 = \frac{2\sqrt{a+p}}{\pi}\int_0^{a+p}\frac{\sigma_o dx}{\sqrt{(a+p)^2 - x^2}} - \frac{2\sqrt{a+p}}{\pi}\int_a^{a+p}\frac{\sigma_Y dx}{\sqrt{(a+p)^2 - x^2}} = 0$$

giving

$$\frac{a}{a+p} = \cos\left(\frac{\pi\sigma_o}{2\sigma_Y}\right) , \tag{32}$$

$$\delta = -\frac{8\sigma_Y a}{\pi E_1}\log\left[\cos\left(\frac{\pi\sigma_o}{2\sigma_Y}\right)\right] . \tag{33}$$

In "plane stress" type of configurations the available results (e.g., [14]) indicate that in the neighborhood of a certain value of the applied load (say the membrane stress σ_o perpendicular to the crack plane), any small increase in σ_o may cause a large increase in the crack opening stretch, δ. This simply means that about this particular applied load the phenomenon which takes place at the leading edge of the crack is very similar to the necking observed in a ductile tensile bar, that is, the material is subjected to "plastic instability". Moreover, after a certain value of σ_o the slopes of the normalized δ vs. σ_o curves increase so rapidly with increasing σ_o that the load corresponding to the plastic instability may be determined from these curves within an acceptable degree of accuracy by assuming a suitably selected standard slope. Once such a critical (standard) slope is selected, δ vs. σ_o curves give the load carrying capacity of the structural element. As an example, Figures 6 and 7 show the application of this criterion to pressurized cylindrical shells containing a through crack. The calculated δ vs. $\sigma_o = N_o/h$ curves are shown in Figure 6, where

$$d_1 = 4a\sigma_Y/E , \quad \lambda = [12(1-\nu^2)]^{1/4}\, a/\sqrt{Rh} , \tag{34}$$

2a is the crack length, R is the (mean) radius, h is the thickness, and σ_Y is the yield strength. The solid line in Figure 7 shows the load carrying capacity of the shell obtained from Figure 6. The data points correspond to the results of the burst tests on various steel pipes as given in [15].

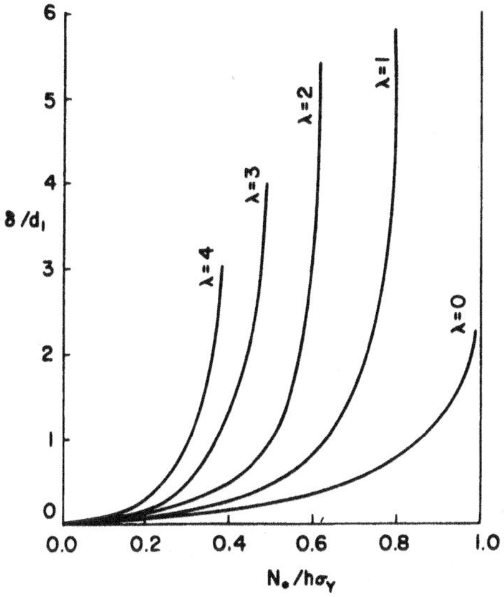

Fig. 6. Crack opening stretch in a pressurized cylindrical shell with a through crack.

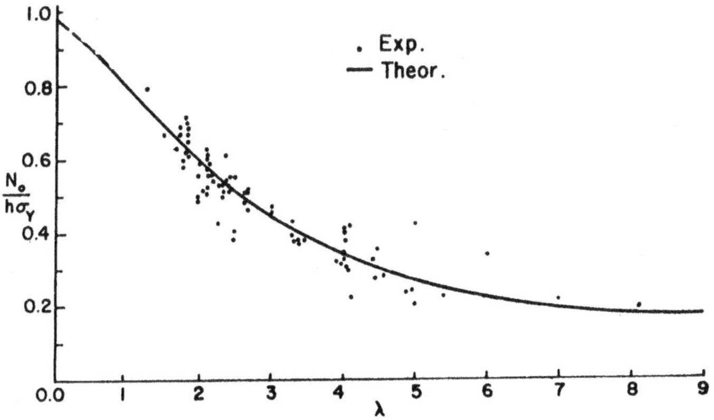

Fig. 7. Comparison of the results of the fracture tests in steel pipes containing a through crack with that given by the plastic instability criterion.

Another fracture criterion of some practical interest, namely the so-called "maximum stress criterion" which appears to be somewhat related to Griffith's later work [16], will be described in the lecture on the Fracture of Nonhomogeneous Materials.

REFERENCES

1. A.A. Griffith, Phil. Trans. Roy. Soc. (London) Ser. A, 221, 163 (1920).
2. E. Orowan, Transactions, Inst. Engrs. Shipbuild., Scotland, 165 (1945).
3. G.R. Irwin, Fracturing of Metals, Am. Soc. Metals, Cleveland, 147 (1948).
4. E. Orowan, Welding Research Supplement, 20, 157s (1955).
5. C.E. Inglis, Transactions, Inst. Naval Architects, 60, 219 (1913).
6. I.N. Sneddon, Fourier Transforms, McGraw-Hill, New York (1951).
7. J.R. Rice, J. Appl. Mech., 34, Trans. ASME, 379 (1968).
8. J.D. Eshelby, Phil. Trans. Roy. Soc. (London) Ser. A, 244, 87 (1951).
9. G.P. Cherepanov, Appl. Math. Mech. (PMM), 31, 503 (1967).
10. Fracture Toughness Evaluation by R-Curve Methods, ASTM, STP527 (1973).
11. A.A. Wells, J. Engng. Fracture Mechanics, 1, 399 (1969).
12. G.D. Fearnehough and B. Watkins, Int. J. Fracture Mechanics, 4, 233 (1968).
13. F. Erdogan and M. Ratwani, Int. J. Fracture Mechanics, 8, 413 (1972).
14. F. Erdogan and M. Ratwani, Proc. Crack Propagation in Pipelines Int. Symposium, Newcastle upon Tyne (1974).
15. J.F. Kiefner, W.A. Maxey, R.J. Eiber, and A.R. Duffy, ASTM Sixth National Symposium on Fracture Mechanics (1972).
16. A.A. Griffith, Proc. First Int. Congress for Appl. Mech., Delft, 55 (1924).

FRACTURE PROBLEMS IN A NONHOMOGENEOUS MEDIUM

F. Erdogan

Lehigh University, Bethlehem, Pa. 18015

1. INTRODUCTION

Most materials one has to deal with in problems relating to geophysics are generally nonhomogeneous. For example, in the lower end of the scale rocks, which are generally assumed to be homogeneous in bulk, usually consist of grains of relatively very high modulus and grain boundary filling material of lower modulus. In the higher end of the scale the layered and stacked big bulks of materials forming the earth's crust may obviously have very different mechanical properties. Thus, from the grain boundary separation in rocks to the propagation of large cracks in the crust running into or going through the interfaces separating two different materials one is basically dealing with fracture of a nonhomogeneous medium.

The main objective of this paper is the discussion of certain characteristic features of the fracture process in nonhomogeneous materials. However, because of the importance of the subject to the fields of rock mechanics and geophysics, some fracture mechanics aspects of problems relating to the failure of materials under compressive loads will first be discussed.

2. FRACTURE MECHANICS ASPECTS OF COMPRESSIVE FAILURE

The fracture theories and criteria discussed in the previous paper [1] have been developed to study the failure of materials which are subjected to mostly "tensile" type of external loads. Since all the working criteria are based on the existence of crack-like flaws and since the cracks close under compression, the extension

Thoft-Christensen (ed.), Continuum Mechanics Aspects of Geodynamics and Rock Fracture Mechanics, 45-64.

of these theories to compressive failure is obviously not straightforward. However, if these theories are physically sound, they should be equally applicable to at least fracture initiation phase of the compressive failure. There is by now sufficient research to indicate that for brittle fracture this is indeed the case [e.g., 2-8].

All the continuum fracture theories start with the conjecture that the material contains a series of randomly oriented and distributed flaws and the macroscopic fracture may initiate at one or simultaneously at a number of these flaws which, for the purpose of analysis, may be treated as plane internal cracks. This also appears to be the case in rocks. For example, in [2] discussing the results of the experiments on rocks Brace concludes that "Grain boundaries at many sites in a rock became loosened as stress was increased prior to fracture. At the instant before fracture a rock was filled with loosened sections of grain boundaries which had various lengths and orientations. Growth of cracks out of these sections of grain boundaries apparently led to fracture." In trying to explain the discrepancy between the theoretical (ideal) strength and the practical (bulk) strength of a solid undergoing brittle fracture and subjected to uniaxial tension, one assumes that the solid contains randomly distributed (penny-shaped) cracks. Usually it is then stated that the tensile strength σ_T is related to (say) k_{IC} through

$$\frac{2}{\pi} \sigma_T \sqrt{a_o} = k_{IC} \tag{1}$$

where a_o is the radius of the (largest) penny-shaped crack and k_{IC} is a material constant which is determined from standard plane strain fracture tests. For this simple case, essentially all theories lead to the same result. With σ_T and k_{IC} known, (1) may also be used to determine the natural flaw size a_o in the material.

Equation (1) (which is obtained from a penny-shaped crack solution for an infinite solid loaded perpendicularly to the plane of the crack) is based on the assumption that the flaw density is sufficiently low so that there is no interaction between the stress field of the dominant crack from which the fracture propagation starts and that of the remaining cracks and the boundaries. Such an interaction would mean that the (calculated) stress intensity factor is usually of the form

$$k = \frac{2}{\pi} \sigma_o \sqrt{a} F(a,d_i) \tag{2}$$

where d_i represents relevant (linear and angular) distances other than the radius a, and the function F is usually an increasing function of a. For example in a plane containing a series of periodically located collinear through cracks the stress intensity factor for loading perpendicular to the plane of the cracks is

$$k = \sigma_o \left(\frac{2b}{\pi} \tan \frac{\pi a}{2b}\right)^{1/2} \tag{3}$$

where 2a is the crack length and 2b is the (periodic) crack distance. The consequence of (2) is that the solid having a higher density of smaller cracks may have the same tensile strength as the one having a lower density of relatively larger cracks. This may be an important consideration particularly in studying compressive fracture.

As pointed out in [1] the fracture process in tension for a through crack or a penny-shaped crack is unstable. Hence, in brittle solids fracture initiating load is also the rupture load. On the other hand in compression tests, the fracture initiating from the most unfavorably oriented crack does not propagate in its own plane and the fracture propagation is almost always stable. This has clearly been shown in [3] on glass specimens containing a dominant through crack. Thus, in studying the failure of (brittle solids under compressive loads a clear distinction must be made between local fracture initiation and the macroscopic rupture along certain fracture surfaces.

The failure of the earlier attempts to apply Griffith's ideas on brittle fracture [1] to explain any phase of the compressive failure of solids such as rocks appears to have been due to neglecting the effects resulting from crack closing under compression. The surfaces of preexisting cracks in the solid are far from smooth frictionless planes; they are generally "jagged" and almost always "rough". The enlarged photographs showing the grain structure of a variety of rocks provide sufficient evidence for this statement [e.g., 6,8]. However, perhaps even more convincing evidence lies in the observation of very short "tensile cracks" initiating from the main crack surfaces in glass specimens subjected to uniaxial compression in which the crack is oriented at an angle to the direction of the load [6,9]. This means that under the action of the compressive stress component perpendicular to the crack, there will be a certain resistance against the relative sliding motion of the crack surfaces, hence, against fracture initiation. This was recognized in [7] and was taken into account by introducing friction on the crack surfaces to oppose the shear component of the applied loads. With this modification, it is now generally agreed that by using the tools of fracture mechanics one could estimate the correct order of magnitude of the fracture initiating load in compressive failure of solids.

To fix the ideas consider the plane problem shown in Figure 1 where it is assumed that $p_1 > p_2 \geq 0$. In terms of the compressive stresses p_1 and p_2 the (compressive) normal and shear components of stresses acting perpendicular and parallel to the crack may be expressed as

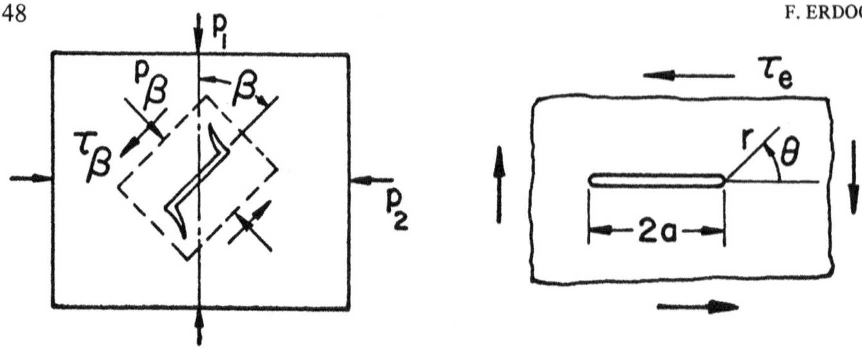

Fig. 1. The geometry of the specimen under compression.

$$p_\beta = \frac{p_1+p_2}{2} - \frac{p_1-p_2}{2}\cos 2\beta ,$$

$$\tau_\beta = \frac{p_1-p_2}{2}\sin 2\beta . \tag{4a,b}$$

Since we are considering a natural crack, the crack surfaces are in contact in the absence of any external loads and if f is assumed to be a constant coefficient of friction, the magnitude of the resisting tangential traction on the crack surface will be fp_β. Thus, in applying the notions of fracture mechanics to the problem the only stress component which will contribute to the singular behavior of the stress state around the crack ends will be the effective shear stress defined by

$$\tau_e = \tau_\beta - fp_\beta . \tag{5}$$

From (4) and (5) it follows that

$$\tau_e = \frac{p_1-p_2}{2}(\sin 2\beta + f\cos 2\beta) - f\,\frac{p_1+p_2}{2} . \tag{6}$$

In the terminology of Figure 1 and (4-6), without any loss in generality it will be assumed that $\tau_e > 0$. The solution of the related elasticity problem gives the stress intensity factors and the stress state in the close neighborhood of the crack tip (on the right-hand side) as follows:

$$k_1 = 0 , \quad k_2 = \tau_e\sqrt{a} , \tag{7a,b}$$

$$\tau_{rr} = \frac{k_2}{\sqrt{2r}}\left[\cos\frac{\theta}{2}\left(2\tan\frac{\theta}{2} - \frac{3}{2}\sin\theta\right) + 0\left(\frac{r}{a}\right)\right] ,$$

$$\tau_{\theta\theta} = \frac{k_2}{\sqrt{2r}}\left[\frac{3}{2}\cos\frac{\theta}{2}\sin\theta + 0\left(\frac{r}{a}\right)\right] ,$$

$$\tau_{r\theta} = \frac{k_2}{2\sqrt{2r}} \left[(1 - 3\cos\theta)\cos\frac{\theta}{2} + O\left(\frac{r}{a}\right)\right] , \qquad (8a\text{-}c)$$

where the polar coordinates r and θ are measured from the crack tip as shown in Figure 1.

First, let us point out that in applying any of the brittle fracture criteria, in this problem k_2 will have to be used as the sole measure of the severity of applied loads and the geometry. Therefore, since the cracks in a real material are randomly oriented, the orientation $\beta = \beta_o$ which will correspond to the maximum value of k_2 for given p_1, p_2, and a, and hence to the crack from which the fracture will initiate, may be obtained from k_2 = maximum, or

$$\frac{\partial \tau_e}{\partial \beta} = 0 , \qquad \frac{\partial^2 \tau_e}{\partial \beta^2} < 0 \qquad (9)$$

giving

$$\tan 2\beta_o = 1/f . \qquad (10)$$

From (10) and Figure 1 one may observe that $0 < \beta_o \leq \pi/4$, $\beta_o \to 0$ as $f \to \infty$, and $\beta_o \to \pi/4$ as $f \to 0$. Substituting from (10) into (6) and (7) we obtain

$$k_2 = \frac{\sqrt{a}}{2} (p_1 \tan\beta_o - p_2 \cot\beta_o) . \qquad (11)$$

Since it is assumed that $p_1 > p_2$ and $\tau_e > 0$, it is clear that (10) and (11) are subject to the following constraint

$$f < \left(\frac{p_1 - p_2}{2} \sin 2\beta_o\right) / \left(\frac{p_1 + p_2}{2} - \frac{p_1 - p_2}{2} \cos 2\beta_o\right) ,$$

or

$$\frac{p_1 - p_2}{p_1 + p_2} > \frac{f}{\sqrt{1 + f^2}} . \qquad (12)$$

The question is now what criterion should one use to determine the fracture initiation load. Since the crack (initially and after applying the loads) is fully closed, in the continuum sense the radius of curvature at the crack tip is zero, the stress concentration factor is infinite, and hence one would have difficulty in using a direct maximum stress criterion based on the elliptic hole solution of Inglis [10,11,7]. Conceptually the most general and perhaps the soundest of all the criteria, namely the energy balance criterion, may also not be applicable in this problem because of the fact that the fracture propagation will not take place in the original plane of the crack and the details of the branching crack problem are not yet available.* If one now goes back to the basic notions of brittle fracture, one may make the following simple hypotheses:

(a) The crack extension starts at the crack tip;

(b) The crack extension will be perpendicular to the direction of local greatest tension;

(c) In homogeneous and isotropic materials at the inception of fracture initiation the magnitude of this local maximum tensile stress (measured at a critical distance) is the same as that prevailing around the tip of a symmetrically loaded cleavage crack.

From hypothesis (b) and (8) the angle of crack initiation θ_o is obtained as

$$\cos\theta_o = 1/3 . \tag{13}$$

If σ_T is the tensile strength of the solid containing a crack of the same size (i.e., $p_1 = -\sigma_T$, $p_2 = 0$, $\beta = 90°$), from the hypothesis (c), i.e., from

$$(\sqrt{2r}\ \tau_{\theta\theta})_T = k_{IC} = \sigma_T\sqrt{a} = (\sqrt{2r}\ \tau_{\theta\theta})_C \tag{14}$$

and from (8), (10), (11) and (13) we obtain

$$p_1 - \frac{p_2}{(\sqrt{1+f^2} - f)^2} = \frac{\sqrt{3}\ \sigma_T}{\sqrt{1+f^2} - f} . \tag{15}$$

In (14) the subscripts T and C stand for tension and compression respectively. Equation (15) gives the fracture initiating compressive stress $p_C = p_1$ in the material for a prescribed lateral pressure p_2 and a specified friction coefficient f.

Even in the ideal plane problem described above, one should remember that the shear tractions along the boundaries of the body are assumed to be zero and the relative size of the crack is small enough to warrant the use of the infinite plane solution. This means that different grip conditions and size effects may change the simple result given by (15) considerably.

In a real test specimen or structural part under compression the ideal plane strain conditions of a single dominant through crack in a relatively large solid are far from being fulfilled. In this

---

* In a recent study Hussain, Pu and Underwood seem to have overcome the analytical difficulty regarding this problem and the complete results may soon be available [12].

case the randomly distributed and oriented cracks are three dimensional having variety of shapes. Even if they are approximated by internal penny-shaped or elliptic cracks, generally the average distance between neighboring flaws may be too small to ignore the interaction of the respective stress fields mentioned previously. In this general three-dimensional case also, through superposition, one may easily reduce the fracture problem to one in which the only external load is a uniform shear traction on each crack surface similar to and similarly obtained as the effective shear stress given by (5). Again due to the crack closure, the tensile (mode I) component of the stress intensity factor k_1 will be zero. However, the forward (mode II) and the antiplane (mode III) shear stress intensity factors k_2 and k_3 will now exist simultaneously along the crack front. For a single penny-shaped or elliptic crack the elastic solution is available (e.g., [13]) from which it is possible to evaluate the compressive stress corresponding to pure mode II or mode III crack initiation. However, the actual process of fracture initiation under combined loading such as this and the crack interaction problem is much too complicated to place any confidence in such calculations. In the plane problem shown in Figure 1 the fracture initiation starts along the entire crack border roughly in the direction predicted by (13) (see, for example the experimental results given in [3] and [6]). On the other hand because of the three-dimensional nature of the crack geometry, in the actual test pieces or structures, the critical condition of fracture is generally reached at one point on the crack border only where the distortion of the crack path will start. Even though a reliable quantitative analysis is difficult to carry out, from what is known of the internal penny-shaped and elliptic crack problems under ideal loading conditions (i.e., comparatively, smaller stress intensity factors than the plane strain values) and from the preceding arguments it is not difficult to conclude that the compressive stress p_1 which will cause fracture initiation in three-dimensional test pieces will be considerably higher than that given by (15) for ideal plane strain specimens.

The other important question of course is about the relationship between fracture initiation stress and the compressive fracture strength of solids. The results of the experiments on the uniaxial compression of glass and photoelastic plates containing a macroscopic crack inclined at a certain angle with respect to the direction of loading indicate that the fracture propagation under compression (unlike that under tension) is not catastrophic, it appears to be stable, and after the propagating crack curves and becomes parallel to the direction of compression it is arrested without any reduction in the magnitude of the applied load [3,6]. This simply means that the compressive fracture strength of the solid is higher than the fracture initiating stress and cannot be directly evaluated from purely fracture mechanics considerations. However, in spite of the self-arresting nature of micro-crack propagation in the three-dimensional solid, it is very likely that this newly formed additional internal fracture surface may sufficiently weaken the shear resistance of the solid in bulk and a mechanism predicted by a macroscopic fracture theory such as Coulomb-Mohr may be responsible for the final fracture. In compressive fracture the phenomena taking place in the solid from the fracture initiation up to the onset of unstable fracture propagation may be qualitatively explained by examining the curves giving the relationship between axial compressive stress and axial, lateral, and volumetric strains for a typical material. Such a relationship for quartzite taken from [6] is shown* in Figure 2. Ignoring the initial small non-linear zone due to the crack closure and the collapse of flat pores, up to a certain pressure p_C (the load level A) all three curves are linear. At p_C the rate of volumetric compression starts decreasing, i.e., the material undergoes dilatancy. This may be interpreted as the initiation of fracture and development of new cracks around the borders of unfavorably oriented internal cracks. Since the ensuing fracture propagation is stable, for further crack propagation the continuous increase in the external load will be necessary. Thus, as the applied load is increased the dilatancy rate will increase due to the further propagation of the growing cracks and/or initiation of new fracture around other less unfavorably oriented initial cracks. At a compressive stress σ_C (the load level B) the rate of

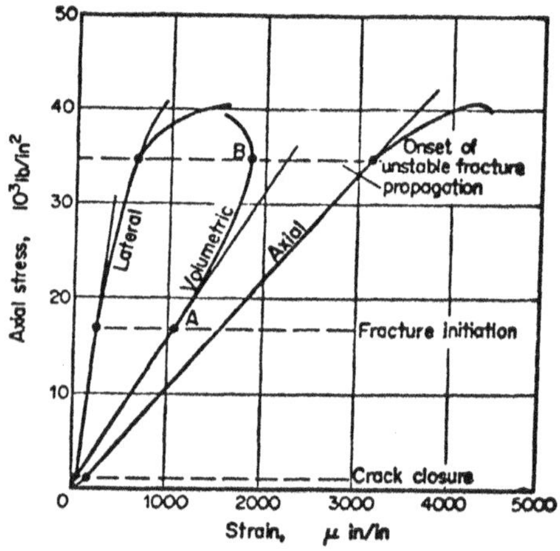

Fig. 2. Axial stress vs. axial, lateral, and volumetric strains in quartzite under uniaxial compression.

* Note that in the compression test the axial (ε_1) and volumetric (ε_v) strains are negative and the lateral strains ($\varepsilon_2 = \varepsilon_3$) are positive and $\varepsilon_v = \varepsilon_1 + \varepsilon_2 + \varepsilon_3$.

change of the volumetric strain becomes infinite implying an instability. It is then clear that this will be the load level at which the unstable fracture propagation will begin. This qualitative discussion indicates that the inability of the so-called fracture mechanics to predict the fracture strength of the solid under combined compressive load conditions from the uniaxial tensile strength is due to the highly complicated (and basically unknown) nature of the geometry of existing cracks and to the intractable nature of the related mathematical problem rather than the physical shortcomings of the relevant fracture theories. In practice the final phase of the fracture process, namely the large scale coalescence of the propagating internal cracks, appears to be adequately treated by using some version of the Coulomb-Mohr theory and the related Mohr envelope which, in its simplest form, may be expressed as

$$|\tau| + \mu\sigma = c \tag{16}$$

where τ and σ are the shear and normal stresses on a given plane, the constant c is known as the cohesive strength and the constant μ is associated with the internal friction. Generally c and μ may be dependent on σ; hence the envelope given by (16) is not necessarily linear.*

3. THE END EFFECTS IN COMPRESSION TESTS

Referring to [14] it appears that in compression tests if the ends of the specimen are laterally constrained (a rough specimen contacting rigid flat ends of the testing machine) the fracture invariably starts at the corners of the specimen (Figures 3a and b). On the other hand in the case of well-lubricated ends (i.e., complete absence of lateral constraints) the macroscopic fracture runs generally along the axial planes. In many rocks lubricated ends cause a reduction of 10 to 20 percent (in Indiana limestone of over 50 percent) in the compressive strength [14]. The increase in the compressive strength due to end constraints may perhaps be explained by arguing that the lateral constraining stresses may have the same effect as confining pressure which is known to improve the fracture strength. The fracture initiation from the corners, however, may be related to the singular behavior of the stresses along the corner of the specimen.

The elasticity solutions for a semi-infinite cylinder and a plane strip with a fixed end are given in [15] and [16] (Figure 4). In this case the stress state around the corners will have the same

---

* For a rather comprehensive discussion of variations of this theory and other macroscopic failure theories see the review article by Paul [14].

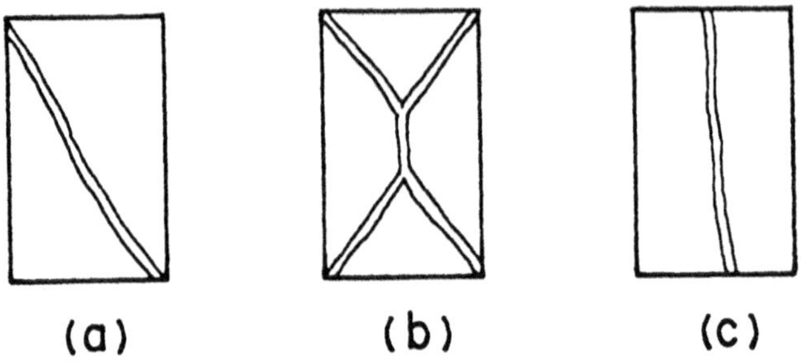

Fig. 3. Fracture patterns in a compressive specimen. (a) and (b) unlubricated, (c) lubricated ends.

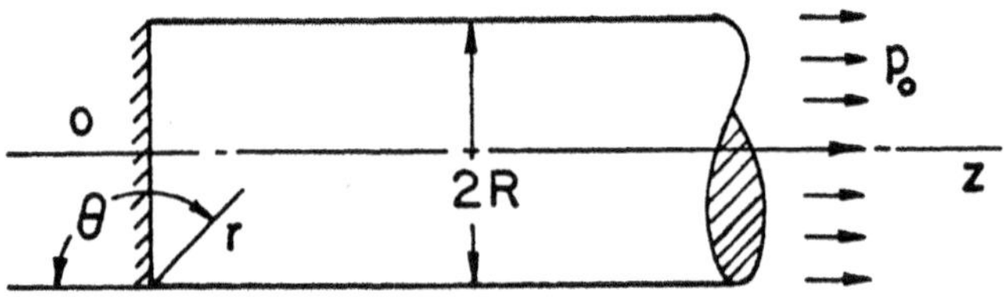

Fig. 4. Cylinder with a fixed end under axial loading.

singular behavior as a crack running into an interface with a rigid half-space (see the following section). For example, in the axisymmetric case the contact stresses σ_{zz} and σ_{rz} will have the following asymptotic behavior:

$$\sigma^T_{zz}(r,0) = \frac{k_1}{\sqrt{2}(R-r')^\alpha} + O[(R-r')^{\alpha_1}] ,$$

$$\sigma^T_{rz}(r,0) = \frac{k_2}{\sqrt{2}(R-r')^\alpha} + O[(R-r')^{\alpha_1}], \quad (Re(\alpha_1) > 0), \quad (17a,b)$$

where R is the radius of the cylinder and the constants k_1 and k_2 are the stress intensity factors. Figures 5 and 6 show the contact stresses for $0 \leq r' < R$ for various values of Poisson's ratio ν. The power of singularity α and the stress intensity factors k_1 and k_2 defined by (17) are shown in Figure 7. Figures 8-10 show the angular distribution of the stresses around the corner expressed in polar coordinates r, θ shown in Figure 4. The fixed grip problem considered in this section (approximately) corresponds to the curves

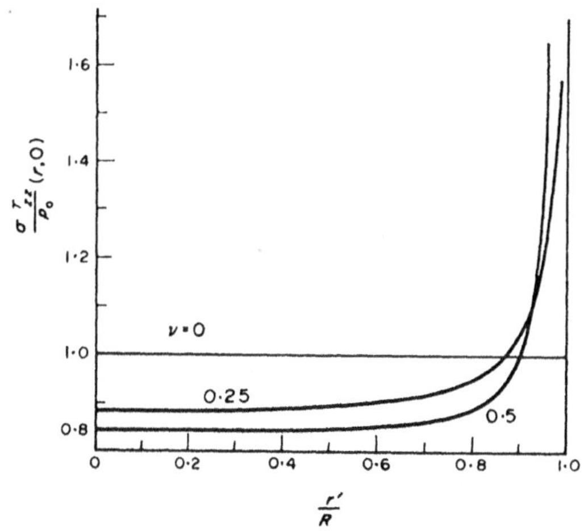

Fig. 5. Normal stress at the fixed end of the cylinder.

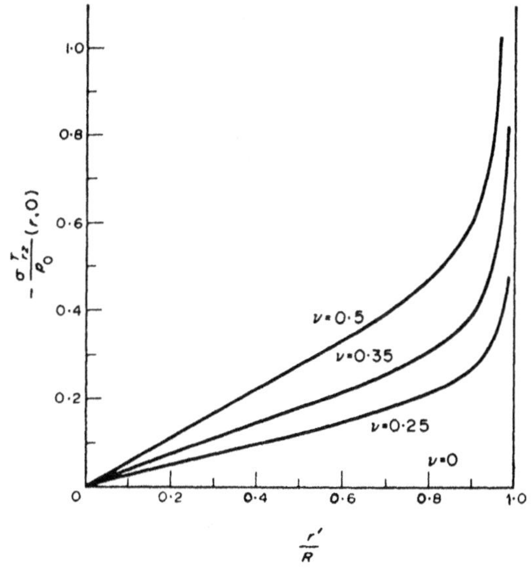

Fig. 6. Shear stress at the fixed end of the cylinder.

given by m = 138.46. It should be noted that the asymptotic stresses around the corner are of the form

$$\sigma_{ij}(r,\theta) = \frac{G_{ij}(\theta)}{\sqrt{2}\, r^{\alpha}} A + O(r^{\alpha_1}), \quad (i,j = r,\theta). \tag{18}$$

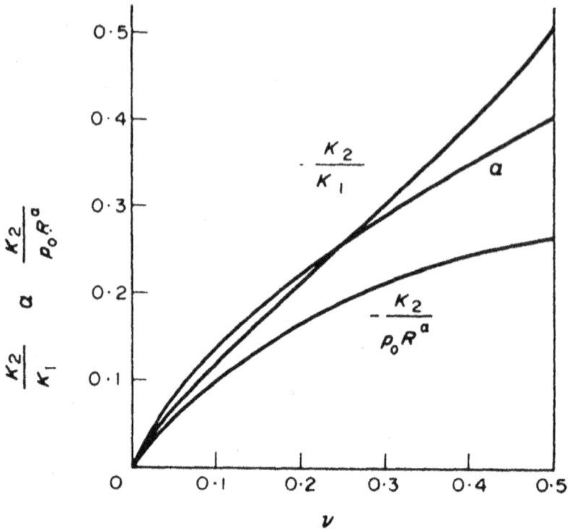

Fig. 7. The stress intensity factors and the power of stress singularity α vs. the Poisson's ratio for a cylinder with fixed end.

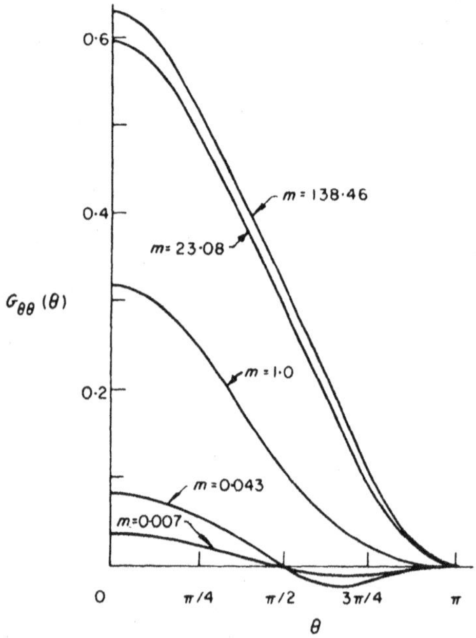

Fig. 8. Angular variation of $\sigma_{\theta\theta}$ around a crack tip touching the interface.

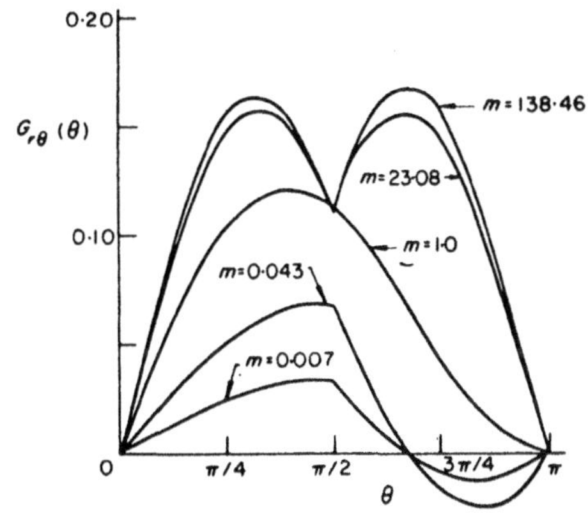

Fig. 9. Angular variation of $\sigma_{r\theta}$ around a crack tip touching the interface.

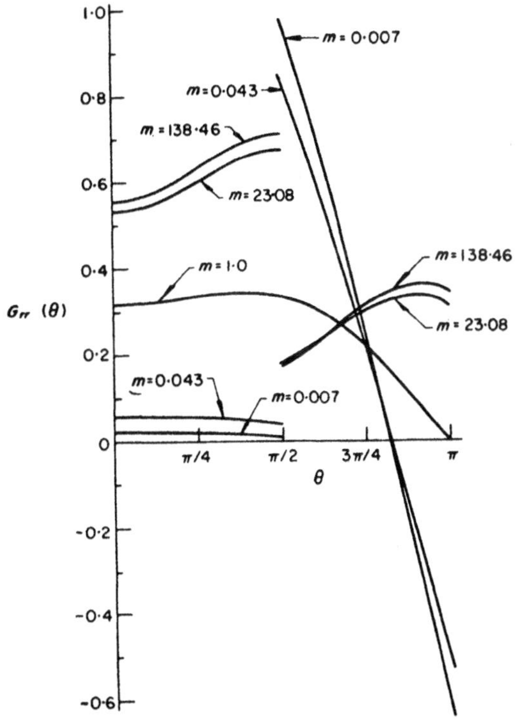

Fig. 10. Angular variation of σ_{rr} around a crack tip touching the interface.

Thus $G_{ij}(\theta)$ gives only the relative distribution (for $\nu = 0.35$), and for quantitative analysis in (18) one needs the value of the multiplying constant A which may be evaluated as

$$A = - \frac{k_2}{G_{r\theta}(\pi/2)} = - 2.11 . \qquad (19)$$

For the applied load $p_o < 0$, $\sigma_{\theta\theta}$ will be negative. Hence, a direct cleavage from the corner is not likely. However, using the technique described in the previous section and the asymptotic results given in this section one may easily study the fracture initiation from the most unfavorably oriented microcrack in the neighborhood of the corner. Whatever the quantitative outcome of such an analysis, it is clear that because of the high stress concentrations around the corners as seen from (18), the fracture initiation and subsequent crack coalescence will start around the corners. The plane strain or the generalized plane stress results similar to those shown in Figures 5-7 are given in [16].

4. FRACTURE OF NONHOMOGENEOUS MATERIALS

If the crack boundary from which the fracture propagation is likely to initiate is fully imbedded in a homogeneous medium, then the techniques described in [1] and in this paper may be used to analyze the fracture problem. However, if the medium is nonhomogeneous in the sense that it consists of perfectly bonded homogeneous constituents and if the crack lies on or intersects a bimaterial interface, then there are stress singularities which are not of the standard $r^{-1/2}$ type, and consequently the fracture criteria described in [1] are either not applicable or need some modifications. In this case the elasticity solutions indicate that the stress state around the crack front is of the following general form [17-21]:

$$\sigma_{ij}(r,\theta) = \frac{1}{r^{\alpha}} [k_1 f_{1ij}(\theta) + k_2 f_{2ij}(\theta)] , \quad (i,j = r,\theta),$$

$$\sigma_{kz}(r,\theta) = \frac{1}{r^{\beta}} k_3 f_{kz}(\theta) , \quad (k = r,\theta),$$

$$\sigma_{zz}(r,\theta) = 0 , \qquad \text{(plane stress)},$$

$$\sigma_{zz}(r,\theta) = \nu(\sigma_{\theta\theta} + \sigma_{rr}) , \quad \text{(plane strain)}. \qquad \text{(20a-d)}$$

where α is generally complex with $0 < \mathrm{Re}(\alpha) < 1$, β is real with $0 < \beta < 1$, the functions $f_{\ell ij}(\theta)$ are bounded, k_1,k_2,k_3 again refer to mode I (tensile), mode II (forward shear), and mode III (antiplane shear) stress intensity factors, and r,θ,z are the cylindrical coordinates measured from the crack front, r,θ plane being normal to the crack border. Two important special cases are the crack on the interface and the crack perpendicular to and intersecting the interface.

For the interface crack the power of singularity is $\alpha = \frac{1}{2} \mp i\omega$, ω being a bimaterial constant. This gives rise to the well-known stress oscillations at the crack, i.e., for small values of r

$$\sigma_{ij} \sim r^{-1/2} \begin{matrix}\cos \\ \sin\end{matrix}[\omega \log(r/a)] \tag{21}$$

where a is the characteristic crack size. For the in-plane loading the stress intensity factors k_1, k_2 may be defined as

$$k_1 + ik_2 = \lim_{x\to a}[2(x-a)]^{1/2}\left(\frac{x-a}{x+a}\right)^{i\omega}[\sigma_{yy}(x,0) + i\sigma_{xy}(x,0)] \tag{22}$$

where 2a is the length of the crack located in the $y=0$ plane. Noting that in this case too as the crack propagates in its own plane the stress state around the crack tip will remain autonomous, the energy available for fracture (for the crack extension on one end) may be evaluated in terms of crack closure energy similar to the derivation of (21) in [1] as follows:

$$\frac{d}{da}(U-V) = \frac{\pi}{2}\frac{1+\kappa_1}{c\mu_1}(k_1^2 + k_2^2) , \tag{23}$$

$$c = \frac{\mu_2 - \mu_1}{\mu_1 + \kappa_1\mu_2} + 2 - \frac{\kappa_2\mu_1 - \kappa_1\mu_2}{\mu_2 + \kappa_2\mu_1}$$

where μ_1, κ_1 and μ_2, κ_2 are respectively the elastic constants of the media occupying $y<0$ and $y>0$, $\kappa = 3-4\nu$ for plane strain, and $\kappa = (3-\nu)/(1+\nu)$ for generalized plane stress. Thus, once the (adhesive) fracture energy, γ_A of the bonding is determined, the energy balance criterion may be used to analyze the fracture problem with (23) as the load factor. For example, in the simple case of two bonded infinite planes with an interface crack of length 2 subjected to tractions $\sigma_{yy}(x,0) = -\sigma_o$, $\sigma_{xy}(x,0) = 0$, $|x| < 1$, the contact stresses and the stress intensity factors may be expressed as [18]

$$\sigma_{xy} - i\sigma_{yy} = -\sigma_o\left(\frac{x-1}{x+1}\right)^{i\omega}\frac{2\omega + ix}{\sqrt{x^2-1}} + i\sigma_o , \quad (y=0,\ x>1),$$

$$k_1 = \sigma_o , \quad k_2 = -2\omega\sigma_o , \quad \omega = \frac{1}{2\pi}\log\left(\frac{1+\gamma}{1-\gamma}\right) ,$$

$$\gamma = \frac{(\mu_1 + \kappa_1\mu_2) - (\mu_2 + \kappa_2\mu_1)}{(\mu_1 + \kappa_1\mu_2) + (\mu_2 + \kappa_2\mu_1)} . \tag{24}$$

Results for other configurations and for layered materials may be found in [18] and in the review article [23].

In nonhomogeneous materials containing a crack which is perpendicular to and terminates at an interface, the power of the singularity α is real, and if the crack is in medium 1 (Figure 11, with $a=0$), the stress intensity factors may be defined as

$$k_1 = \lim_{r\to 0}\sqrt{2}\, r^{\alpha}\sigma_{2\theta\theta}(r,0)\,, \quad k_2 = \lim_{r\to 0}\sqrt{2}\, r^{\alpha}\sigma_{2r\theta}(r,0)\,. \tag{25a,b}$$

For the symmetric problem the asymptotic values of the stresses are

$$\sigma_{ij}(r,\theta) \simeq \frac{k_1}{\sqrt{2}r^{\alpha}}\,\frac{G_{ij}(\theta)}{G_{\theta\theta}(0)}\,, \quad (i,j=r,\theta) \tag{26}$$

where the functions $G_{ij}(\theta)$ are shown in Figures 8-10 for various material combinations*. Table 1 gives an idea about the variation of α and the stress intensity factors. Here the material 1 is a matrix containing a crack of length 2a which terminates radially at the interface of a circular inclusion (material 2) the radius of which is $R = 2a$, $\kappa_1 = \kappa_2 = 1.8$, the matrix is uniaxially loaded perpendicular to the crack ($\sigma_{yy}^{\infty} = \sigma_0$), $k(a)$ is the stress intensity factor at the interface crack tip (as defined by (25a)), and $k(b)$ is the stress intensity factor at the other end.

If the crack goes through the interface, there are three singular points, b_1, b_2, and 0 (see insert in Figure 12). At the ends b_1 and b_2 the stress intensity factors may be defined and the

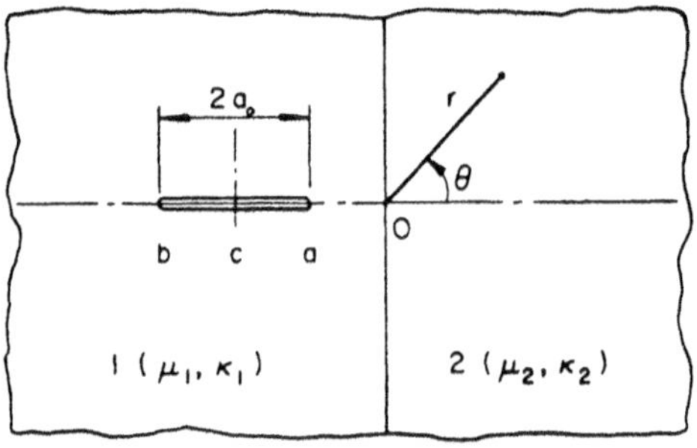

Fig. 11. Geometry for a crack perpendicular to the interface.

* In descending order of $m = \mu_2/\mu_1$ the material pairs are epoxy (1)-boron (2), epoxy-aluminum, homogeneous medium, aluminum-epoxy, and boron-epoxy.

Table 1. Stress intensity factors for a crack terminating at the interface of a matrix-inclusion interface.

$\frac{\mu_2}{\mu_1}$	α	$\frac{k(b)}{\sigma_o\sqrt{a}}$	$\frac{k(a)}{\sigma_o a^\alpha}$
0		1.940	
0.05	0.81730	1.615	1.053
1/3	0.62049	1.229	0.5836
1.0	0.5	1.000	1.000
3.0	0.40074	0.8610	1.299
10.0	0.33277	0.7969	1.389
23.0	0.30959	0.7796	1.375
100	0.29387	0.7691	1.345
300	0.28883	0.7667	1.348

fracture propagation problem may be treated in a conventional way. At the interface $r = 0$ the normal and shear cleavage components of the stress intensity factors may be defined as

$$k_\theta = \lim_{r\to 0} r^\alpha \sigma_{1\theta\theta}(r,\tfrac{\pi}{2}), \quad k_r = \lim_{r\to 0} r^\alpha \sigma_{1r\theta}(r,\tfrac{\pi}{2}), \tag{27a,b}$$

where α is real and $0 < \alpha < 0.5$. Figures 12 and 13 show the results of an example. Further results may be found in [19-23].

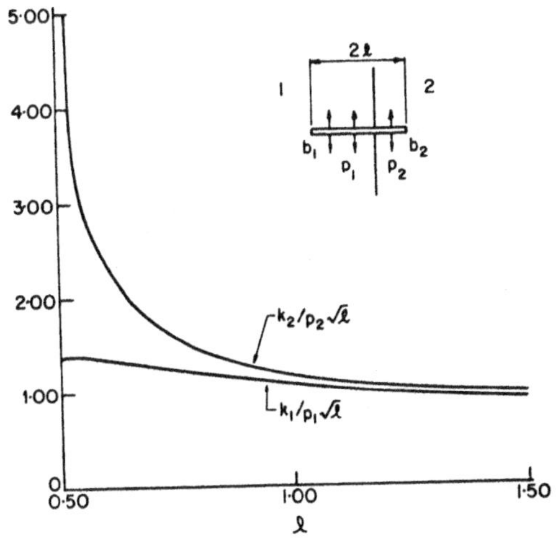

Fig. 12. Stress intensity factors for a crack crossing the interface (Materials: 1: Aluminum, 2: Epoxy; $b_1 = 1 =$ constant, load: crack surface pressures p_1, p_2, $p_1/p_2 = E_1/E_2$).

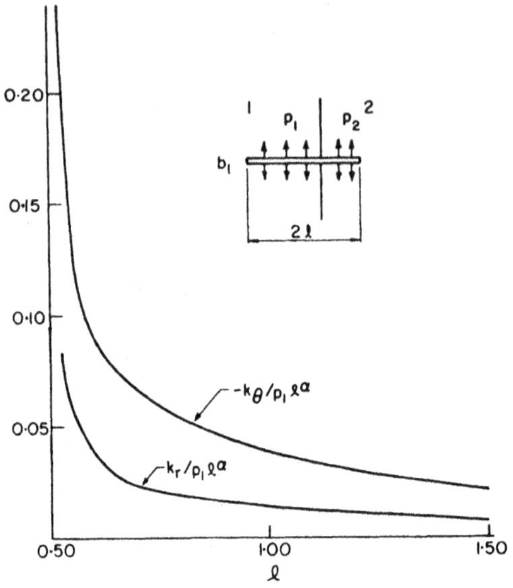

Fig. 13. Stress intensity factors for cleavage and shear stresses on the interface for a crack crossing the interface (Materials: 1: Aluminum, 2: Epoxy; $b_1 = 1 =$ constant, load: crack surface pressures p_1, p_2, $p_1/p_2 = E_1/E_2$; power of singularity: $\alpha = 0.27369$).

For crack propagation initiating from the singular point at the interface, since the stress state around the crack tip does not remain self-similar or autonomous as the crack propagates, most of the conventional fracture theories do not seem to be applicable. At present, in the absence of a physically more acceptable criterion, the maximum cleavage stress criterion may serve the purpose for brittle materials. This is essentially the concept described in section 2 of this paper and states that

The fracture propagation will take place in the direction $\theta = \theta_c$ for which the cleavage stress $\sigma_{\theta\theta}$ is maximum and when

$$\sigma_{\theta\theta}(\delta_p, \theta_c) = \sigma_c \ . \tag{28}$$

In (28) δ_p is the size of the fracture process zone and depends on the microstructure and the continuum properties of the material, and on the environmental conditions. In brittle materials it is sufficiently small so that the response of the medium for $r > \delta_p$ is not significantly affected by what takes place inside this zone. This means that for most applications an elastic analysis may be sufficient. σ_c is a constant representing the cohesive strength of the constituent materials. If the "weak link" around the singular point is the interface, then (28) may be modified to read

$$(\sigma_c)_{12} = \begin{cases} (\sigma_{\theta\theta}^2 + \sigma_{r\theta}^2)^{1/2}, & \sigma_{\theta\theta} > 0, \\ \sigma_{r\theta} + f\sigma_{\theta\theta}, & \sigma_{\theta\theta} < 0, \end{cases} \tag{29}$$

where $(\sigma_c)_{12}$ now represents the adhesive strength of the bond 1-2, $f\sigma_{\theta\theta}$ represents the friction resistance, and the stresses $\sigma_{\theta\theta}$, $\sigma_{r\theta}$ are calculated at $r = (\delta_p)_{12}$ representing the fracture process zone size for the particular joint. The constants σ_c and $(\sigma_c)_{12}$ may be determined from idealized laboratory experiments once a reasonable estimate for δ_p and $(\delta_p)_{12}$ is made. With the asymptotic nature of the stress state around the singular point known, as for example given by (26), the application of this criterion is quite straightforward. It should be pointed out that this criterion may also be applied if the singular point is the apex of a wedge rather than a crack tip. For example in a plane wedge of (material) angle $2\theta_o$ the asymptotic stress state around the apex will be of the form

$$\sigma_{ij}(r,\theta) \simeq \frac{k}{r^\alpha} f_{ij}(\theta) ,$$

$$(0 \leq \theta < \theta_o, \ \frac{\pi}{2} \leq \theta_o \leq \pi, \ 0 \leq \alpha \leq \frac{1}{2}), \tag{30}$$

which is similar to the stress fields found for the crack problems. This particular problem may have some applications in rock fracture during drilling operations.

REFERENCES

1. F. Erdogan, "Principles of Fracture Mechanics", This Volume.
2. W.F. Brace, State of Stress in Earth's Crust, ed. W. Judd, Elsevier, New York (1964).
3. W.F. Brace and E.G. Bombolakis, Journal of Geophysical Research, 68, 3709 (1963).
4. W.F. Brace, Journal of Geophysical Research, 65, 4377 (1960).
5. Z.T. Bieniawski, Int. J. Rock Mech. Min. Sci., 4, 395 (1967).
6. Z.T. Bieniawski, Int. J. Rock Mech. Min. Sci., 4, 407 (1967).
7. F.A. McClintock and J.B. Walsh, Proc. 4th U.S. Nat. Congress of Appl. Mech., 2, 1015 (1962).
8. W.F. Brace, E. Silver, K. Hadley, and C. Goetze, Science, 178, 162 (1972).
9. E. Sommer (private communication).
10. C.E. Inglis, Transactions, Inst. Naval Architects, 60, 219 (1913).
11. A.A. Griffith, Proc. 1st Int. Congress for Appl. Mech., Delft, (1924).
12. M.A. Hussain, S.L. Pu, and J. Underwood, U.S. Army Watervliet Arsenal Report (1974).
13. G.C. Sih and H. Liebowitz, Fracture, 2, ed. H. Liebowitz, 67, Academic Press, New York (1968).

14. B. Paul, Fracture, 2, ed. H. Liebowitz, 313, Academic Press, New York (1968).
15. G.D. Gupta, Int. J. Solids Structures, 10, 137 (1974).
16. G.D. Gupta, J. Appl. Mech., 40, Trans. ASME, 948 (1973).
17. F. Erdogan, J. Appl. Mech., 30, Trans. ASME, 232 (1963).
18. F. Erdogan and G.D. Gupta, Int. J. Solids Structures, 7, 1089 (1971).
19. T.S. Cook and F. Erdogan, Int. J. Engng. Sci., 10, 677 (1972).
20. F. Erdogan and V. Biricikoglu, Int. J. Engng. Sci., 11, 745 (1973).
21. F. Erdogan, Fracture Mechanics of Ceramics, 1, ed. R.C. Bradt, D.P.H. Hasselman, and F.F. Lange, 245, Plenum Press, New York-London (1974).
22. F. Erdogan, J. Engineering Fracture Mechanics, 4, 811 (1972).

DYNAMICS OF FRACTURE PROPAGATION

F. Erdogan

Lehigh University, Bethlehem, Pa. 18015

1. INTRODUCTION

In discussing the fracture of solids under a single application of the load, generally one may differentiate three types of material response. The first is the so-called crystalline shatter of a defect-free crystal with an ideally uniform geometry and under ideally uniform external loads, or fragmentation of defect-free high strength brittle solids. In the case of perfect crystals the fracture may be a complete shatter of the material or the instantaneous rupture along a plane. In this case too the initial fracture may be followed by material fragmentation which will be discussed briefly later in this paper. The second type of response may arise in another ideal case in which the geometry and the microstructure of the medium and the external loads are such that the fracture nucleation and growth (i.e., the formation and propagation of microcracks) take place uniformly and simultaneously along a certain plane. In such a case the fracture propagation from each microcrack may have similar velocities. Even though these velocities may in themselves be somewhat limited, due to the multiplicity of fracture nuclei, the final phase of the fracture, i.e., the process of rapid crack propagation, may be of very short duration. A carefully grooved homogeneous thin sheet under uniform tension may come close to satisfying the conditions necessary for this type of fracture.

The third and the much more realistic and common material response is the onset of fracture propagation from a dominant flaw upon the magnitude of the applied loads reaching a critical value. For some materials, in this case also, one may observe the formation and coalescence of cracks and voids in the solid. However,

Thoft-Christensen (ed.), Continuum Mechanics Aspects of Geodynamics and Rock Fracture Mechanics, 65-79.

the small cracks or voids in question here are formed ahead of the main propagating crack and are due to the stress concentration caused by the main crack. The terms "crack propagation" or "fracture propagation" in current literature are used almost exclusively in connection with this third type of material response. In this paper therefore we will be interested mainly in discussing the dynamic aspects of the fracture propagation in solids. In particular we will be concerned with a certain group of problems in which, because of the high rates of loading and/or high crack velocities, the inertia effects in solving the field equations and the strain rate and the velocity effects in the strength characterization of the material are not negligible.

2. TENSILE FRAGMENTATION OF FIBERS

Even though the fracture problems one encounters in practice generally fall in the category of crack propagation from a dominant flaw, from the physical viewpoint, particularly in connection with the dynamic fracture of solids, the tensile fragmentation of high strength fibers is sufficiently important to warrant a brief discussion. The phenomenon has been observed in very thin (e.g., 30 μm.) defect-free fused silica or virgin glass fibers and in alumina whiskers subjected essentially to static tensile loads [1]. Studies relating to the dynamics of this fragmentation process by means of high speed photography indicate that [1] the high strength fiber fixed between two grips and subjected to tension does not break into two pieces as in ordinary tensile tests but simply disintegrates. In [1] the successive pictures were taken at a rate of 3 to 4×10^5 frames per second. In the glass experiments, the size of the fragments were estimated to be 0.1 micron. The results of successive pictures corresponding to periodic intervals are sketched in Figure 1. The remarkable aspect of these experiments was that

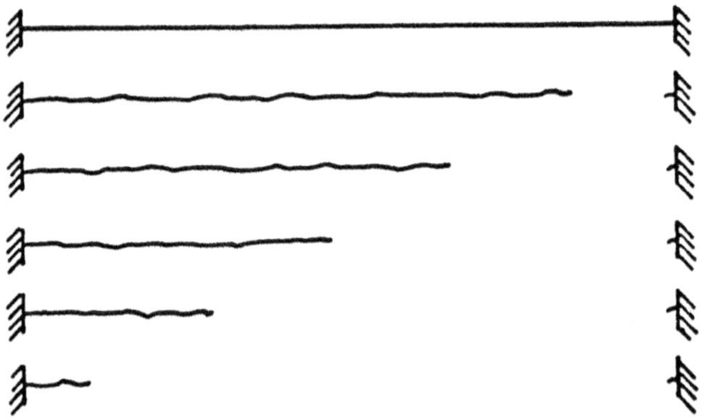

Fig. 1. Disintegration of a high strength wire under tension.

in all cases it was found that the residual fiber length versus time relationship was linear, and for a given material its slope was constant. This implies that for each material the fragmentation velocity was constant (in glass, approximately 1.8×10^4 cm/sec).

In the continuum model (e.g., [2] and [3]) one assumes that, upon rupture, two types of stress waves travel along the bar, namely, the longitudinal waves and the flexural waves (caused by the radially antisymmetric nature of the initial rupture). Trying to explain the fragmentation as being the consequence of repeatedly reflected and superimposed stress waves does not seem to be possible. This is because the stress wave approach would predict dependence of the disintegration velocity on time and a strong likelihood of rupture closer to the fixed end of the fiber. Both of these predictions are contrary to the experimental observations. A possible explanation of the phenomenon may lie in the approach of wave mechanics [4], where it is conjectured that the destruction of the bonds may be caused by the field-free particles which are the group of atoms accelerated in the solid due to the dynamic nature of the applied loads.

3. REFLECTED STRESS WAVES IN SLENDER BARS

The earliest experimental work regarding the dynamic effects on the rupture strength of materials was carried out by J. Hopkinson [5]. He performed drop-weight rupture tests on steel wires which were thick enough to contain the randomly distributed defects normally associated with the macroscopic bulk of the material. Thus, in this case the dominant flaw theory of fracture in conjunction with the reflected stress waves should be valid for the explanation and the analysis of the results. In the drop-weight experiment if M is the mass of the falling weight, m the mass of the retaining clamp at the end of the bar, h the height from which the mass is dropped, ρ the mass density of the bar, g the gravitational constant, A the cross-sectional area of the bar, the stress at a distance x from the impacted end of the bar may be expressed as [6,7]

$$\sigma = \frac{\rho c M}{M+m} (2gh)^{1/2} \exp\left[\frac{\rho A c}{M+m} (x-ct)\right], \quad (x \leq ct), \qquad (1)$$

where $c = (E/\rho)^{1/2}$ is the velocity of the longitudinal waves in the bar and t the time measured from the instant of impact. Since the tension pulse will be doubled in magnitude at the fixed end [7] (see Figure 2a), in long bars it is expected that the fracture would take place at the fixed end after the first reflection. However, as shown by Taylor [8], it is possible to observe higher stresses at the fixed end (hence, fracture at a later time) because of the superposition of the reflected pulse from the end having the weight. After the first reflection the highest stress at the fixed end is

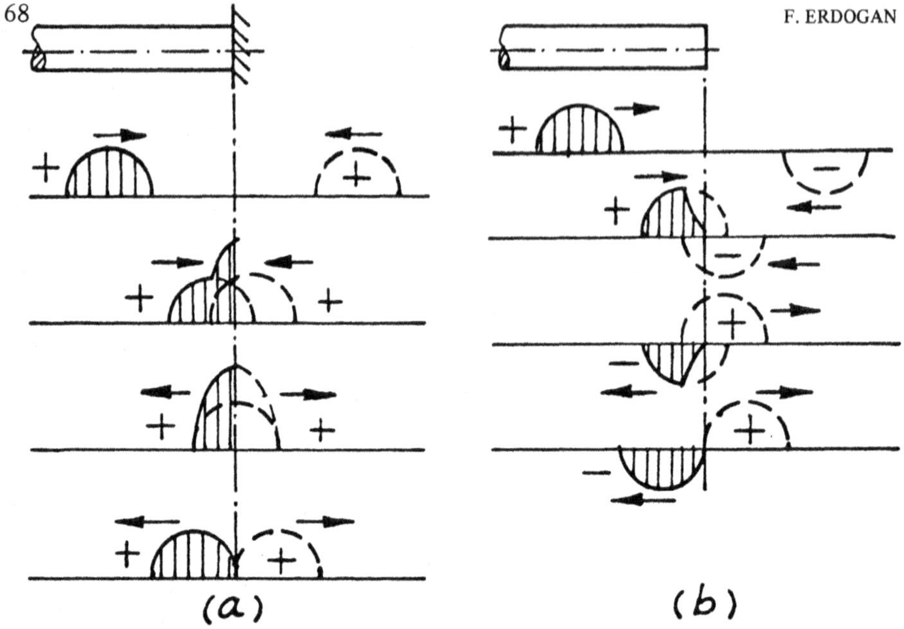

Fig. 2. Reflection of a stress pulse from a fixed end (a), and from a free end (b).

$$\sigma = 2\rho c V_o, \qquad V_o = \frac{M}{m+M} (2gh)^{1/2}, \tag{2}$$

where V_o is the particle velocity at $t = 0$, $x = 0$. In B. Hopkinson's experiments [9] the maximum stress was calculated by Taylor [8] to be 2.15 $\rho c V_o$ which occurred after the second reflection at the fixed end.

The reflection of pulses from a free end of the bar is shown in Figure 2b [7]. Here, the stress at the end is always zero, and an incoming tension wave is reflected as a compression wave (or vice versa). Basically this is the cause of the so-called spall-fracture produced by compression waves [9]. In a preliminary study the assumption of nondispersive linear theory described by Figure 2 may be adequate. However, at levels of stress causing fracture it should be remembered that the material response is no longer linear, and dispersion due to the stress-free lateral surface may change the pulse profile significantly. In studies regarding the fracture of "bars" and spall-fracture in plates, one may assume that the fracture will spread from a dominant flaw with the most unfavorable orientation when the magnitude of the tensile stress exceeds a certain critical limit. From the viewpoint of fracture propagation, the resulting crack configurations are generally unstable. Therefore, a sustained crack growth may be realized by a decreasing load magnitude. For example, in the Hopkinson bar, despite the exponential decrease in the stress amplitude given in equation (1), the bar ruptures completely at the fixed end. However, in spallation experiments [9], it is possible to have the first reflected wave produce a complete rupture, and the second reflected wave produce only an internal fracture in which the crack is arrested because of the decrease in stress level. This indicates the importance of the initial time profile of the pulse in fracture experiments. Depending on the fracture propagation velocity in a given material and configuration, to produce fracture the profile of the incoming pulse must therefore be such that a certain level of stress can be maintained for a certain duration of time.

4. DIFFRACTION BY A STATIONARY CRACK

In the simple fracture experiments described in the previous section as well as in other more general dynamic fracture problems, the dynamic effects may be discussed as two separate problems. One is the determination of the level of the external load producing the stress waves which will start the fracture propagation from a dominant stationary flaw. This is essentially a problem of calculating a "dynamic stress concentration" and characterizing the material for fracture initiation under impact type of loading. The second problem (which arises if the crack starts to propagate) is the dynamics of the propagating crack under a sustained applied stress of certain time-dependence and the related material characterization regarding the dynamic nature of the fracture resistance. In this section we will discuss briefly the dynamic aspects of the problem for a stationary crack. In practice the problem may arise when a crack of certain size suddenly appears in (or is introduced into) a stressed solid, or stress waves of certain time profile generated elsewhere in the solid due to external impact loading pass through the region of the solid containing an initial internal crack. The main analytical problem here is the solution of the related field equations and the evaluation of the quantities (such as the stress intensity factors) which are used in the application of the corresponding fracture criterion. The general problem for solids with finite dimensions with an arbitrary internal crack appears to be at the present time mathematically intractable. The problems which have been studied are the plane and axisymmetric problems for an infinite linearly elastic solid containing a single crack.

The relatively general plane strain problem of diffraction of a plane dilatational wave (P-wave) (Figure 3a) and a plane vertically polarized shear wave (SV-wave) (Figure 3b) by a finite crack was considered in [10]. The same problem for a horizontally polarized shear wave (SH-wave) (the anti-plane shear problem) was considered in [11] and [12]. In these problems it is assumed that a plane wave of certain time profile traveling at the appropriate

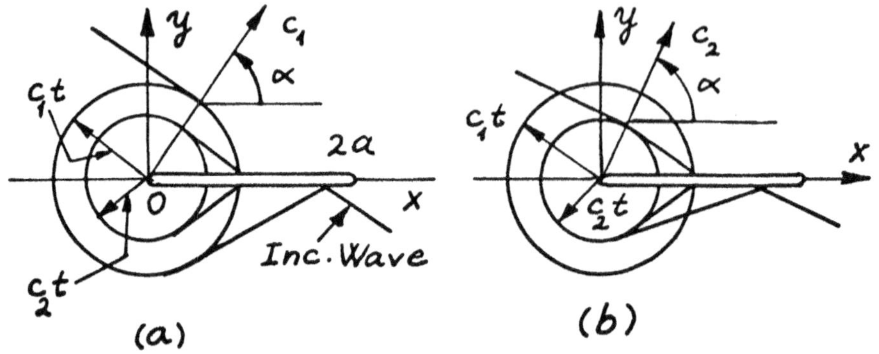

Fig. 3. A dilatational (a) and a vertically polarized shear (b) stress wave striking a finite crack.

wave velocity (c_1 for the dilatational waves and c_2 for the shear waves) strikes a stationary crack and is diffracted, where

$$c_1 = [(\lambda+2\mu)/\rho]^{1/2}, \quad c_2 = (\mu/\rho)^{1/2}, \tag{3}$$

λ and μ being the Lamé's constants which are related to engineering elastic constants by

$$\mu = \frac{E}{2(1+\nu)}, \quad \lambda = \frac{E\nu}{(1+\nu)(1-2\nu)}. \tag{4}$$

Even though the complete solution of these problems is rather complicated, the qualitative description of the resulting wave fronts is quite straightforward. For example, in plane problems, taking $t=0$ as the instant the wave front strikes the first crack tip $x=0$, $y=0$, for $0<t<2a/c_1$ the crack will generate two reflected waves and two cylindrical diffracted waves centered at the crack tip, with some shadow regions developing as shown in Figure 3. At $t=2a/c_1$ the diffraction will start from the second crack tip and as time increases the diffraction of cylindrical waves will continue. In studying the special case of the problem for $\alpha=\pi/2$ for plane strain [13] and for a penny-shaped crack [14] it was shown that in the dynamic problem the stress state around the crack tip may be expressed as

$$\sigma_{ij}(r,\theta,t) \simeq \frac{1}{\sqrt{2r}}\,[A_1(t)k_1^s f_{1ij}(\theta) + A_2(t)k_2^s f_{2ij}(\theta)],$$

$$f_{1ij}(\theta) = f_{1ij}(-\theta), \quad f_{2ij}(\theta) = -f_{2ij}(-\theta), \quad (i,j=r,\theta), \tag{5}$$

where (r,θ) are the polar coordinates at the crack tip, k_1^s and k_2^s are the corresponding normal (i.e., symmetric) and in-plane shear components of the stress intensity factor, the functions $A_1(t)$ and $A_2(t)$ are the dynamic stress intensity correction factors, and the functions f_{1ij} and f_{2ij} are identical to those obtained for the static problem. A similar expression exists for the anti-plane shear problems:

$$\sigma_{zj}(r,\theta) \approx \frac{1}{\sqrt{2r}} A_3(t)k_3^s f_{3j}(\theta), \quad (j = r,\theta), \tag{6}$$

where f_{3j} is again the same as in the static case.

In symmetric plane strain [15] and in anti-plane shear [16] problems the asymptotic values of the functions $A_1(t)$ and $A_2(t)$ for small values of t are shown to be

$$A_1(t) \approx 0.785\sqrt{c_2t/a}, \quad (t \ll 2a/c_1,\ c_2/c_1 = 0.542),$$

$$A_3(t) \approx \frac{2\sqrt{2}}{\pi}\sqrt{c_2t/a}, \quad (t \ll 2a/c_2). \tag{7a,b}$$

In [10] and [11,12] the solutions for plane strain and anti-plane shear problems were obtained exactly for the time intervals of $0 < t < 4a/c_1$ and $0 < t < 4a/c_2$, respectively. These solutions not only verify the parabolic time dependence of the functions A_i (i.e., $A_i(t) \sim \sqrt{t}$) for $t \ll a/c_2$, they give the exact values of $(A_i)_{max}$ (i.e., the dynamic overshoot) and their time of occurrence. For a step-function incident wave it was shown that*

$$[A_1(t)]_{max} \approx 1.30, \quad [A_2(t)]_{max} \approx 1.20,$$

$$[A_3(t)]_{max} = \frac{4}{\pi}, \tag{8a-c}$$

and the time of occurrence of the peak stress intensities was the instant of arrival of the (first) scattered Rayleigh wave from the opposite crack tip in plane strain problems and $4a/c_2$ in the anti-plane shear problems. The value given in (8) for A_1 appears to be independent of the angle of incidence α. There is a slight dependence on α in $(A_2)_{max}$ (1.20 is for $\alpha = 60°$; for $\alpha = 90°$ the value is 1.18). The exact expression for A_3 at the crack tip $x = 0$ is [12]

$$A_3(t) = \frac{2}{\pi}\left[\frac{c_2t}{a(1+\cos\alpha)}\right]^{1/2} \sin\alpha, \quad (0 \le t \le 4a/c_2). \tag{9}$$

The solution of the axisymmetric impact problem for a solid with a penny-shaped crack given in [14] indicates that $(A_1)_{max}$ in this case is probably identical to that of plane P-waves (i.e., 1.30).

It should be pointed out that the dynamic overshoot mentioned

* In [10] the Poisson's ratio was assumed to be 1/4 in calculating A_1 and A_2.

above is valid only if there is a stress discontinuity at the wave front. Thus, if the plane wave has a continuous front, the loading around the crack tip will not be a step function in time, and consequently, depending on the wave profile, the overshoots may be considerably smaller than those given by (8).

Assuming that the "overshoot" discussed in this section gives some idea about the dynamic effects on the critical load factor representing the severity of the applied loads, in fracture studies one also needs to know the similar effects on the fracture resistance parameter of the material. The kind of experimental work this may require in which the dynamic effects can be properly isolated and measured does not appear to have yet been carried out. It may be conjectured that these effects would be more significant in the strain-rate sensitive materials, and in general since the toughness of most structural materials decreases with increasing loading rates, under the impact type of loading considered in this section the fracture resistance of the material would be somewhat lower than the corresponding static value. In recent years there is a tendency toward defining a "dynamic fracture toughness" G_{Id} (or corresponding k_{Id}) and characterizing the material by means of the results of pre-fatigue-cracked Charpy impact tests (as opposed to G_{Ic} characterization which is done under "slow" loading). As a basis of comparison of the dynamic fracture resistances of two different materials, tests such as Charpy impact, or drop-weight-beam, or drop-weight-tear may be perfectly adequate. However, for a quantitative study, they are clearly not suitable. In these as well as in the fracture propagation problems which will be discussed in the next section, the main culprit appears to be the irreversible effects (such as plasticity) which, even in static problems, have not yet been properly modeled.

5. FRACTURE PROPAGATION

From the practical viewpoint perhaps the most important problem regarding the dynamic effects is that of fracture propagation. The general problem is the following: A deformable solid with a given geometry is subjected to a system of time-dependent external loads, generally consisting of surface tractions T_i, surface displacements u_i, and body forces F_i, $(i = 1,2,3)$, and contains an initial (dominant) flaw which serves as a fracture nucleus. If the external loads are increased beyond a critical level, the initial flaw (which is usually an internal or a surface crack) will start to propagate. The related mechanics problem is then the determination of all the desired field quantities (such as stresses and displacements) and the size and shape of the fracture area as a function of time, including the asymptotic states leading to fracture arrest or catastrophic failure. At this generality even the formulation of the problem does not seem to be feasible. The dynamic problems which can be formulated are those in which the fracture area A(t) can be characterized in terms of a single length parameter a(t) [17], and the problems which have been solved (in an inverse manner as will be described in this section) are the plane or the axisymmetric problems for linearly elastic unbounded solids. In this case there are four unknown functions, u_1, u_2, u_3, and a(t). Three equations are provided by*

$$\mu\nabla^2 u_i + (\lambda+\mu)\frac{\partial e}{\partial x_i} + F_i = \rho\,\frac{\partial^2 u_i}{\partial t^2}\,, \qquad e = \sum_1^3 \frac{\partial u_i}{\partial x_i}\,, \qquad (i = 1,2,3), \qquad (10)$$

where λ and μ are the Lamé's constants and ρ is the mass density. Equations (10) will be subject to appropriate initial and boundary conditions. The fourth equation to account for the additional unknown function, a(t), will have to be provided by some kind of a fracture criterion. Since the propagating crack remains in its own plane, an appropriate and a very general criterion would be that of energy balance, which may for the whole body be expressed as [18]

$$\frac{dU}{dt} = \frac{dV}{dt} + \frac{dK}{dt} + \frac{dD}{dt}\,, \qquad (11)$$

where U is the work done by the "external loads" (including the body forces), V the internal energy, K the kinetic energy, and D the energy dissipated due to fracture propagation. For a solid with volume R and surface S, the first three quantities may easily be expressed as

$$\frac{dU}{dt} = \int_S T_i^n \dot{u}_i\, dS + \int_R F_i \dot{u}_i\, dR\,,$$

$$\frac{dV}{dt} = \int_R \sigma_{ij}\dot{\varepsilon}_{ij}\, dR\,,$$

$$\frac{dK}{dt} = \int_R \rho\dot{u}_i\ddot{u}_i\, dR\,, \qquad (12a\text{-}c)$$

where the dot indicates differentiation with respect to time and the summation convention is used. Since

$$S(t) = S_o + A(t)\,, \qquad (13)$$

A(t) being the fracture surface, defining a specific fracture

---

* Here it is assumed that the material is basically elastic in the sense that the size of the inelastic zone around the crack border is too small to have any significant effect on the overall stress distribution and on the flow of energy into the fracture process zone at the crack tip (see [17]).

energy γ_F (i.e., energy required to create a unit amount of fracture surface) by [17]

$$\frac{dD}{dA} = \gamma_F \tag{14}$$

the fourth term in (11) becomes

$$\frac{dD}{dt} = \gamma_F \frac{dA}{dt} , \quad A = f(a) . \tag{15}$$

The fracture energy γ_F is generally a function of the crack velocity $\dot{a}$ and strain rates in the dissipation zone. Even for highly brittle materials there is very little useful information available on γ_F. There is some experimental evidence [e.g., 19] which indicates that as $\dot{a}$ increases γ_F first decreases, goes through a plateau region, and then increases rather rapidly. However, since none of these experiments lend themselves to any type of a rigorous analysis, the results are highly qualitative. There appears to be no work done with regard to isolating and studying the effect of strain rates in the fracture process zone on γ_F. Partly for these reasons, and mostly because of the highly nonlinear nature of the problem (see equations 10-15), up to now no attempt has been made to solve equations (10) and (11) directly. The solutions which are available on the fracture propagation problems have been obtained in an indirect way in which the fracture criterion (11) is replaced by the fact that the characteristic crack size a is a known function of time (see the review articles [18] and [20] for the recent developments). After solving the problem, presumably one may go back to (11) for verification.

Since one is usually dealing with an unbounded medium, evaluation of the terms in (11) may appear to be the source of another major difficulty. However, as shown in [18], in the type of "linear" problems considered in this paper, for the application of the energy balance criterion the global evaluation of the terms in (11) is not necessary and the required information may be obtained from the asymptotic solution valid in the close neighborhood of the propagating crack tip (see also [20] and [21]). The result found in [18] was

$$\frac{dU}{dt} - \frac{dK}{dt} - \frac{dV}{dt} = \frac{dE_c}{dt} = \frac{dD}{dt} , \tag{16}$$

where E_c is the crack closure energy for unit increase in a and may be obtained from (see Figure 4)

$$dE_c = \frac{dE_c}{dt} dt = 2L \int_a^{a+\dot{a}dt} \frac{1}{2} [\sigma_{\theta\theta}(s)u_\theta(s - \dot{a}dt)$$

$$+ \sigma_{r\theta}(s)u_r(s - \dot{a}dt)]ds , \tag{17}$$

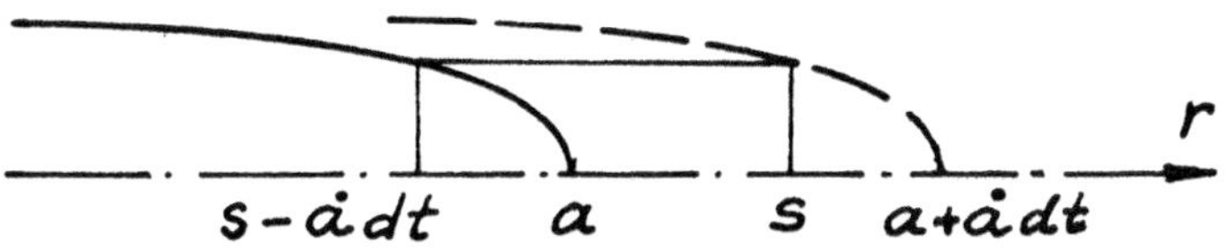

Fig. 4. Notation for calculation of crack closure energy rate $\dot{E}_c$.

where r and θ are the polar coordinates in the plane perpendicular to the crack border, L is the length of the crack periphery at a(t) (i.e., $L = 2\pi a$ for a penny-shaped crack, $L = h$ for a crack growing from one end, and $L = 2h$ for a crack growing at both ends with the same velocity $\dot{a}$, h being the thickness of the medium in plane problems). Thus, if the asymptotic solution for the cleavage and shear stresses $\sigma_{\theta\theta}$ and $\sigma_{r\theta}$ and for the corresponding displacements u_θ and u_r is available around the crack tip, $\dot{E}_c$ which, according to (16), represents the energy available for fracture at the crack front may easily be found from (17), and (16) and (15) would then give the fracture criterion.

There are limited number of solutions for a semi-infinite crack propagating at a nonuniform velocity ([16] and [22] for anti-plane shear and [23] for plane strain problems). However, in most of the solutions of the crack propagation problems which appeared in literature it is assumed that $\dot{a}$ is constant. An up-to-date review of the existing solutions may be found in [18] and [20], and therefore will not be attempted in this paper. Instead we will discuss only some of the significant results.

Consider an infinite "plane" which is initially at rest and is loaded uniaxially by a uniform tension $\sigma_{yy}^{\infty} = p$ at infinity. Assume that at $t = 0$ a through crack appears and spreads in $y = 0$ plane at a constant velocity (at both ends) $\dot{a}$, the total crack length being $2a = 2\dot{a}t$ [24,25]. In this case at a time t the cleavage stress $\sigma_{yy}(x,0,t)$, ($x = \dot{a}t+r$, $r \ll \dot{a}t$) ahead of the crack may be expressed as

$$\sigma_{yy}(\dot{a}t+r,0,t) \simeq \frac{k_d(t)}{\sqrt{2r}} = p(c_2 t/2r)^{1/2} F(v) ,$$

$$v = \dot{a}/c_2 , \qquad (18a,b)$$

where k_d is the dynamic stress intensity factor, c_2 is the shear wave velocity (see 3) and the expression for F(v) may be found in [18] or [24]. At time t the corresponding static stress intensity factor is

$$k_s = p\sqrt{a} = p\sqrt{c_2 vt} . \qquad (19)$$

From (18) and (19) it follows that

$$k_d = p\sqrt{c_2 t}\ F(v)\ , \quad \frac{k_d}{k_s} = \frac{F(v)}{\sqrt{v}}\ . \tag{20}$$

Figure 5 shows the variation of k_d and k_s with the velocity ratio v normalized with respect to $p\sqrt{c_2 t}$, (i.e., F(v) and $\sqrt{v}$). At $v = c_R/c_2$, F(v) and consequently k_d become zero where c_R is the Rayleigh wave velocity (in this example, $\nu = 1/4$, $c_R = 0.9194\, c_2$).

Similarly, using the asymptotic results given in [24] one may also evaluate the rate of the energy available for fracture (per unit thickness) at the crack tip, $\dot{E}_c$, as follows:

$$\frac{dE_c}{dt} = \frac{\pi p^2 c_2^2 t v^4}{\mu H^2(v)}\ [4(1-v^2)^{1/2} - \frac{(2-v^2)^{1/2}}{(1-k^2v^2)^{1/2}}]\ ,$$

$$H(v) = 8E(v_1) - [4 + (2-v^2)^2/v_2^2]E(v_2) - 4v^2K(v_1)$$

$$+ [v^4 + k^2v^2(2-v^2)^2/v_2^2]K(v_2)\ ,$$

$$k = c_2/c_1\ , \quad v_1 = (1-v^2)^{1/2}, \quad v_2 = (1-k^2v^2)^{1/2}. \tag{21}$$

In limit as $v \to 0$ (i.e., the static case), $H(v) \to -2v^2(1-k^2)$ and from (21) it is found that

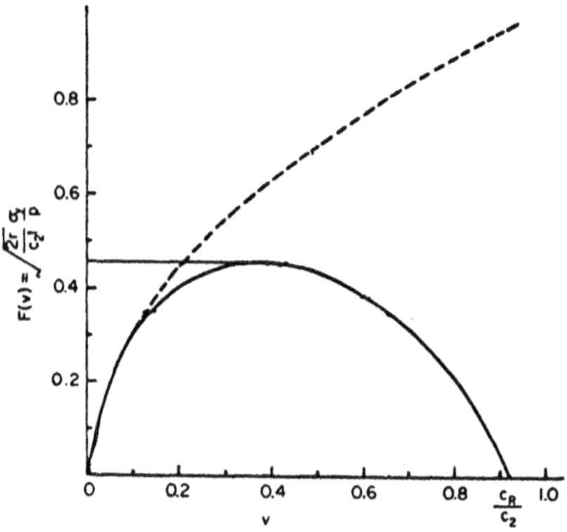

Fig. 5. Measures of dynamic (——) and static (---) stress intensity factors versus velocity ratio $v = \dot{a}/c_2$ in a crack propagating at constant velocity at both ends.

$$\frac{1}{vc_2}\frac{dE_c}{dt} = \frac{dE_c}{da} = \frac{\pi}{2}\frac{pc_2vt}{\mu(1-k^2)} = \frac{2\pi}{E}p^2a(1-\nu^2) , \tag{22}$$

which is the known result. Noting that the Rayleigh wave velocity c_R is a root of (see, for example, [7])

$$16(1 - k^2c_R^2/c_2^2)(1 - k^2) - (2 - c_R^2/c_2^2)^4 = 0 , \tag{23}$$

from (21) it is seen that at $v = c_R/c_2$ (or at $\dot{a} = c_R$) $\dot{E}_c$ will be zero. The obvious consequence of this result is that, provided the crack can be constrained to remain in its own plane, c_R is an upper limit for the velocity of a propagating crack. Remembering the meaning of $\dot{E}_c$ as being the energy available for fracture at the crack tip, this also means that to have fracture velocities greater than c_R, energy must be generated (rather than dissipated) at the crack front. Even though this may not be possible in ordinary structural applications, super-Rayleigh fracture velocities have been observed in the fracture of certain transparent dielectrics by a laser pulse [26,27]. In these experiments it is believed that the additional energy is provided by the gas resulting from thermal destruction of the material at the crack tip (see also [28] on the wedging of a solid by super-Rayleigh velocity).

For the plane problem under consideration the fracture area (per unit thickness) A is 4a, and (15) and (16) give the energy balance equation as

$$\dot{E}_c = \dot{D} = \gamma_F\dot{A} = 4\gamma_F\dot{a} = 4\gamma_F vc_2 . \tag{24}$$

On the other hand from (21) it is seen that $\dot{E}_c$, the energy available for fracture, increases with time linearly. Thus, if the fracture energy γ_F is constant or does not increase with increasing crack length (i.e., with time and/or crack velocity) sufficiently fast, there will be a constantly increasing excess energy rate $\dot{E}_c - \dot{D}$ which will accelerate the crack, under ideal conditions causing its velocity to approach asymptotically the Rayleigh wave velocity. In practice this of course never happens; either the crack branches in order to absorb the excess energy, or the crack velocity reaches a "terminal" constant value and stays there because of increasing fracture resistance with crack velocity. For example, if one assumes that the dissipation is primarily due to plastic work which is proportional to the volume of the plastic zone, for unit thickness one may write $D = Ba^2 = Bc_2^2v^2t^2$ or $\dot{D} = 2Bc_2^2v^2t$, B being a constant. In this case the fracture criterion $\dot{E}_c = \dot{D}$ would not contain t and would provide an equation to determine the equilibrium crack velocity, provided the constant B can be accurately estimated. This simple notion is described in Figure 6 where $\dot{E}_c$ is obtained from (21). From Figure 6 it is seen that for $v < v_T$, $\dot{E}-\dot{D} > 0$, there is excess energy, and the crack would accelerate

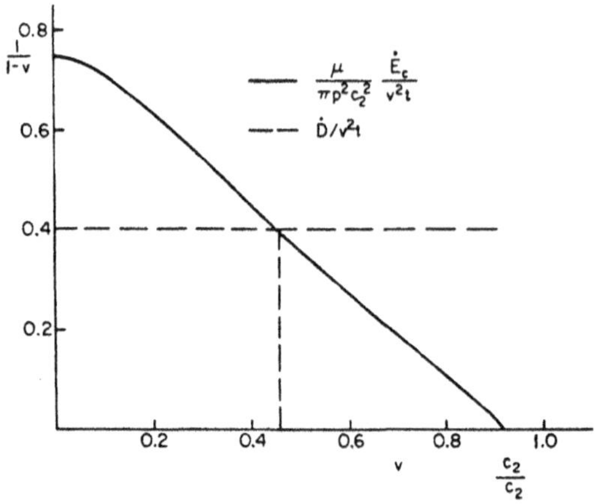

Fig. 6. Energy balance for propagating crack in a solid for which $D = Ba^2$.

(i.e., v tends to increase). Also if $v > v_T$, $\dot{E} - \dot{D} < 0$ and the crack would decelerate. The velocity $v_T c_2 = \dot{a}_T$ then corresponds to the terminal crack velocity.

There is a substantial amount of experimental work on the dynamic effects on fracture of solids, and a great variety of experimental techniques have been developed to study the problem. Some of them have been referred to in this paper [1, 5, 6, 9, 19, 26, 27], and most of them had been reviewed in [6] and [18]. However no paper on the subject could be complete without acknowledging the outstanding work of Schardin, Kerkhof, Sommer, and their colleagues on glass and hard polymers which has been summarized in a recent book [29].

REFERENCES

1. S. Hyodo and M. Kimura, Fracture Mechanics of Ceramics, 1, R.C. Bradt, D.P.H. Hasselman, and F.F. Lange, eds., 431, Plenum Press, New York (1974).
2. J. Miklowitz, J. Appl. Mech., ASME, 20, 122 (1953).
3. J.W. Phillips, Int. J. Solids Structures, 6, 1403 (1970).
4. E.R. Fitzgerald, Particle Waves and Deformation in Crystalline Solids, Interscience Publ., New York (1966).
5. J. Hopkinson, Original Papers, 2, B. Hopkinson, ed., 316, Cambridge Univ. Press (1901).
6. H. Kolsky and D. Rader, Fracture, 1, H. Liebowitz, ed., 533 Academic Press, New York (1968).
7. S. Timoshenko and J.N. Goodier, Theory of Elasticity, McGraw-Hill, New York (1951).

8. G.I. Taylor, J. Inst. Civil Engrs., 26, 486 (1946).
9. B. Hopkinson, Proc. Roy. Soc. (London), Ser. A, 74, 498 (1905).
10. S.A. Thau and T.H. Lu, Int. J. Solids Structures, 7, 731 (1971).
11. S.A. Thau and T.H. Lu, Int. J. Engng. Sci., 8, 857 (1970).
12. J.D. Achenbach, Int. J. Engng. Sci., 8, 947 (1970).
13. G.C. Sih, G.T. Embley, and R.S. Ravera, Int. J. Solids Structures, 8, 977 (1972).
14. G.T. Embley and G.C. Sih, Developments in Mechanics, 6, Proc. 12th Midwestern Mechanics Conf., 473 (1971).
15. B.R. Baker, J. Appl. Mech., ASME, 29, 449 (1962).
16. B.V. Kostrov, PMM (J. Appl. Math. Mech.), 30, 1241 (1966).
17. F. Erdogan, Principles of Fracture Mechanics, Article in this Volume.
18. F. Erdogan, Fracture, 2, H. Liebowitz, ed., 498, Academic Press, New York (1968).
19. T.L. Paxson and R.A. Lucas, Proc. Symposium on Dynamic Fracture, Lehigh University (1972).
20. J.D. Achenbach, Mechanics Today, 1, S. Nemat-Nasser, ed., 1, Pergamon (1972).
21. L.B. Freund, J. of Elasticity, 2, 341 (1972).
22. J.D. Eshelby, J. Mech. Phys. Solids, 17, 179 (1969).
23. L.B. Freund, J. Mech. Phys. Solids, 20, 141 (1972).
24. K.B. Broberg, Arkiv F. Fysik, 18, 159 (1960).
25. J.W. Craggs, Fracture of Solids, D.C. Drucker and J.J. Gilman, eds., J. Wiley, New York (1963).
26. G.I. Barenblatt, N.N. Vsevolodov, L.I. Mirkin, N.F. Pilipetsky, and Yu.P. Raiser, Zh ETP Letters, V, 3 (1967).
27. S. Winkler, D.A. Shockey, and D.R. Curran, Int. J. Fracture Mechanics, 6, 2 (1970).
28. G.I. Barenblatt and R.V. Goldstein, Int. J. Fracture Mechanics, 8, 427 (1972).
29. F. Kerkhof, Bruchvorgänge in Gläsern, Verlag der Deutschen Glastechnischen Gesellschaft (1970).

NONLOCAL ELASTICITY AND WAVES

A. Cemal Eringen

Princeton University
Princeton, New Jersey 08540

ABSTRACT. The theory of nonlocal elasticity is developed. Balance laws, jump conditions and the second law of thermodynamics are given. By means of axioms of nonlocality, objectivity and the entropy inequality, the constitutive equations are derived for nonlocal elastic solids. The field equations are obtained and applied to study the propagation of body and surface waves.

1. INTRODUCTION

The inner structure of materials and the intermolecular long range forces give rise to a certain class of physical phenomena that cannot be explained by means of classical continuum mechanics. For example, according to classical elasticity, plane waves propagating in an elastic solid are nondispersive, i.e., the phase velocity is independent of the wave length; in fact it is a constant. The same situation is valid for the Rayleigh surface waves. Phonon dispersion experiments abundantly available in the literature (cf. Brockhouse [1962], Harrison [1963], Wallis [1963]) have shown that the phase velocity depends strongly on the wave length, especially in the short wave length region. This situation and a plethora of other phenomena (e.g., surface tension, surface energy, optical branches of the dispersion curve, etc.) cannot be explained by means of the classical elasticity.

In fact, it is not necessary to address the atomic nature of materials for such discrepancies. For granular materials (e.g.,

---

* This work was partially supported by the Army Research Office-Durham.

Thoft-Christensen (ed.), Continuum Mechanics Aspects of Geodynamics and Rock Fracture Mechanics, 81-105.

rocks) and fiber composites, the same situation arises when the wave length becomes comparable to the average size of grains or the average distance between them. Thus, in the study of propagation of shocks and earthquake waves we face similar difficulties.

The purpose of the present paper is to give an exposition of the nonlocal elasticity with applications to wave propagations, in order to show how this new theory can account for these phenomena correctly for wave lengths comparable all the way to the atomic distances.

The theory of nonlocal elasticity is of recent origin. Early works (although to some extent incomplete--no thermodynamics--and heuristic) are due to Kröner [1967], Kunin [1968], Eringen [1967] and Edelen [1969]. Recently we have constructed a nonlocal thermodynamics and rational constitutive theory, on par with the axiomatic approach of contemporary continuum mechanics, Eringen [1972a], [1974a]. A variational approach for nondissipative cases (elasticity) was also provided by Edelen and Laws [1971], see also Eringen and Edelen [1972b]. The theories of polar, nonlocal continua, nonlocal fluid dynamics, nonlocal microfluid mechanics, nonlocal E-M elastic solids, nonlocal continuum thermodynamics, nonlocal thermoelasticity, nonlocal memory dependent materials, etc., were also developed and applied to the solution of some problems, Eringen [1972a,d], [1973a,b,c,d], [1974a,b,c]. In the present work we give yet another approach to the constitutive theory, with tighter postulates leading to concise results both in the theory and its applications. The predictions of very simple nonlocal moduli, which make the dispersion curves coincident with those of the atomic lattice theory in the *entire* Brillouin zone, and the expression of energy which predicts frequency square dependence are the main results of the theory.

In section 2 and 3 we establish the nonlocal balance laws and the entropy inequality. In section 4 we develop the constitutive theory for the nonlinear nonlocal elasticity and linearize the results in section 5. In section 6 the field equations are obtained, and it is shown that the surface physics (surface stresses) is included in the theory. Section 7 deals with the propagation of plane waves in isotropic solids. The nonlocal elastic moduli are determined, and the expression of strain energy per wave number is obtained. In section 8 we treat the problem of surface waves and show that the dispersion curves are coincident with that of the atomic lattice theory in the entire Brillouin zone.

While the theory of nonlocal elasticity is rather young, its power and potential appears to be great in dealing with static and dynamic problems of solids in which the nature of the internal structure of a solid and long range interatomic forces are important.

2. BALANCE LAWS

In the reference state a material point of the body B is identified by the rectangular coordinates X^K, (K = 1,2,3) or simply by $\underset{\sim}{X}$ from an origin $\underset{\sim}{O}$. The motion carries $\underset{\sim}{X}$ to a spatial position $\underset{\sim}{x}$ identified by its rectangular coordinates x^k, at time t. The motion is a one-parameter mapping

$$x^k = x^k(X^K, t) \qquad \text{or } \underset{\sim}{x} = \underset{\sim}{x}(\underset{\sim}{X},t) \tag{2.1}$$

which is assumed to possess a unique inverse

$$X^K = X^K(x^k, t) \qquad \text{or } \underset{\sim}{X} = \underset{\sim}{X}(\underset{\sim}{x},t) \tag{2.2}$$

throughout the body B. Thus, we assume

$$\det x^k_{,K} > 0 \qquad \text{in B} \tag{2.3}$$

Throughout this paper indices following a comma indicate partial differentiations with respect to coordinates and a superposed dot or D/Dt denote material time derivative, e.g.,

$$x^k_{,K} \equiv \partial x^k/\partial X^K \quad , \quad X^K_{,k} \equiv \partial X^K/\partial x^k$$

$$\dot{v}^k \equiv Dv^k/Dt \equiv \partial v^k(\underset{\sim}{x},t)/\partial t\Big|_{\underset{\sim}{X}} = \frac{\partial v^k}{\partial t} + v^k_{,\ell}\dot{x}^\ell$$

where and throughout repeated indices indicate summation over the range (1,2,3).

The integral balance laws of continuum mechanics may be expressed in the general form

$$\frac{d}{dt}\int_{V-\sigma} \phi \, dv - \oint_{S-\sigma} \tau^k da_k - \int_{V-\sigma} g \, dv = 0 \tag{2.4}$$

where ϕ is *a field* over the body $\mathcal{B}$ at time t having material volume V excluding a discontinuity surface σ which may be sweeping the body by a velocity $\underset{\sim}{u}$, g is *the source* of ϕ and τ^k is its *influx* through the surface ϕ of V excluding those points of σ intersecting S.

By means of the generalized Green-Gauss theorem,(2.4) may be converted into (cf. Eringen [1967, p. 77]:

$$\int_{V-\sigma} [\frac{\partial\phi}{\partial t} + (\phi v^k)_{,k} - \tau^k_{,k} - g]dv + \int_\sigma [\phi(v^k - u^k) - \tau^k]n_k da = 0 \tag{2.5}$$

where $v^k \equiv \dot{x}^k$ is the velocity field in $V-\sigma$ and a boldface bracket $[\ ]$ indicates the *jump* across $\sigma(t)$.

In classical continuum mechanics it is posited that (2.5) is valid for every part of the body, and, therefore, integrands of (2.5) are set equal to zero. As a result, one obtains local laws of continuum mechanics. In nonlocal continuum physics we reject this severe assumption. Localization may still be accomplished by writing (2.5) in the equivalent form

$$\begin{aligned} &\frac{\partial\phi}{\partial t} + (\phi v^k)_{,k} - \tau^k_{,k} - g = \hat{g} && \text{in } V - \sigma \\ &[\underset{\sim}{\phi}(v^k - u^k) - \tau^k - \underset{\sim}{\hat{G}^k}]n_k = 0 && \text{on } \sigma \end{aligned} \tag{2.6}$$

such that

$$\int_{V-\sigma} \hat{g}\, dv + \int_\sigma [\underset{\sim}{\hat{G}^k}] d\underset{\sim}{a_k} = 0 \tag{2.7}$$

The new fields $\hat{g}$ and $\hat{G}^k$ introduced are called "localization residuals" or "nonlocal interactions". Determination of the nonlocal interaction is, of course, an integral part of the nonlocal continuum physics.

Equations (2.6) and (2.7) are the *master balance laws* of nonlocal continuum mechanics. We now employ these equations to obtain special balance equations for mass, momentum, moment of momentum and energy.

(i) *Mass.* If we take $\phi = \rho$, $\tau^k = g = 0$ where ρ is the mass density in $V - \sigma$ then (2.6) gives

$$\begin{aligned} &\frac{\partial\rho}{\partial t} + (\rho v^k)_{,k} = \hat{\rho} && \text{in } V - \sigma \\ &[\underset{\sim}{\rho}(v^k - u^k) - \underset{\sim}{\hat{R}^k}]n_k = 0 && \text{on } \sigma \\ &\int_{V-\sigma} \hat{\rho}\, dv + \int_\sigma [\underset{\sim}{\hat{R}^k}] d\underset{\sim}{a_k} = 0 \end{aligned} \tag{2.8}$$

where $\hat{\rho}$ and $\hat{R}^k$ are the nonlocal mass interactions in $V-\sigma$ and on σ respectively.

(ii) *Momentum.* By setting $\phi = \rho\underset{\sim}{v}$, $\tau^k = \underset{\sim}{t}^k$ and $g = \rho\underset{\sim}{f}$ in (2.6), we obtain the nonlocal balance law for momentum

$$\begin{aligned} &\underset{\sim}{t}^k_{,k} + \rho(\underset{\sim}{f} - \dot{\underset{\sim}{v}}) = \hat{\rho}\underset{\sim}{v} - \rho\hat{\underset{\sim}{f}} && \text{in } V - \sigma \\ &[\rho\underset{\sim}{v}(v^k - u^k) - \underset{\sim}{t}^k - \underset{\sim}{\hat{F}^k}] da_k = 0 && \text{on } \sigma \\ &\int_{V-\sigma} \rho\hat{\underset{\sim}{f}}\, dv + \int_\sigma [\underset{\sim}{\hat{F}^k}] d\underset{\sim}{a_k} = 0 \end{aligned} \tag{2.9}$$

where $\rho\hat{\underset{\sim}{f}}$ and $\hat{\underset{\sim}{F}}^k$ are the nonlocal force interactions in $V-\sigma$ and on σ, respectively. Here $\underset{\sim}{t}^k$ is the stress vector and $\underset{\sim}{f}$ is the body force per unit mass. In terms of the stress tensor $t^{k\ell}$ we have

$$\underset{\sim}{t}^k = t^{k\ell}\underset{\sim}{i}_\ell \tag{2.10}$$

where $\underset{\sim}{i}_\ell$ are unit cartesian vectors in the spatial rectangular frame of reference x^k.

(iii) *Moment of Momentum.* To obtain the nonlocal balance law for the moment of momentum in (2.6) we set

$$\phi = \underset{\sim}{x} \times \rho\underset{\sim}{v} \quad , \quad \tau^k = \underset{\sim}{x} \times \underset{\sim}{t}^k \quad , \quad \underset{\sim}{g} = \underset{\sim}{x} \times \rho\underset{\sim}{f}$$

$$\hat{\underset{\sim}{g}} = \underset{\sim}{x} \times \rho\hat{\underset{\sim}{f}} \qquad \hat{G}^k = \underset{\sim}{x} \times \hat{\underset{\sim}{F}}^k$$

and use (2.8) and (2.9). This gives

$$\underset{\sim}{i}_k \times \underset{\sim}{t}^k = 0 \qquad \text{or} \qquad t^{k\ell} = t^{\ell k}$$

$$\int_{V-\sigma} \underset{\sim}{x} \times \rho\hat{\underset{\sim}{f}}\, dv + \int_\sigma \underset{\sim}{x} \times \hat{\underset{\sim}{F}}^k\, da_k = 0 \tag{2.11}$$

Thus, the *stress tensor is symmetric.* Note that this classical result is due to a specific choice of nonlocal moment of momentum residuals, $\hat{g}$ and $\hat{G}^k$, in the form as given above. By the same token and due to the fact that $[\underset{\sim}{x}] = 0$ (i.e., motion is assumed to be continuous across σ), the jump condition associated with (2.11) is satisfied identically. This choice of $\hat{g}$ and $\hat{G}^k$ is certainly in harmony with the forms of ϕ, τ^k and g and appears to be reasonable on physical grounds.

(iv) *Energy.* The nonlocal law of energy balance is obtained if in (2.6) we set

$$\phi = \rho\varepsilon + \frac{1}{2}\rho\underset{\sim}{v} \cdot \underset{\sim}{v} \;, \quad \tau^k = \underset{\sim}{t}^k\cdot\underset{\sim}{v} + q^k \;, \quad g = \rho\underset{\sim}{f}\cdot\underset{\sim}{v} + \rho h$$

$$\hat{g} = \rho\hat{h} \quad , \quad \hat{G}^k = \hat{H}^k$$

and use (2.8) and (2.9). Here ε is the internal energy density per unit mass, q^k is the surface energy influx (e.g., the heat vector) and h is the energy source per unit mass. Thus

$$\rho\dot{\varepsilon} - \underset{\sim}{t}^k\cdot\underset{\sim}{v}_{,k} - q^k_{,k} - \rho h + \hat{\rho}(\varepsilon - \frac{1}{2}\underset{\sim}{v}\cdot\underset{\sim}{v}) + \rho\hat{\underset{\sim}{f}}\cdot\underset{\sim}{v} = \rho\hat{h} \quad \text{in } V-\sigma$$

$$[(\rho\varepsilon + \frac{1}{2}\underset{\sim}{v}\cdot\underset{\sim}{v})(v^k - u^k) - \underset{\sim}{t}^k\cdot\underset{\sim}{v} - q^k - \hat{H}^k]n_k = 0 \quad \text{on } \sigma \tag{2.12}$$

$$\int_{V-\sigma} \rho\hat{h}\, dv + \int_\sigma [\hat{H}^k]\, da_k = 0$$

Equations (2.8), (2.9), (2.11) and (2.12) are the fundamental balance laws of nonlocal continuum mechanics. They are valid for all types of bodies (fluids, solids, viscoelastic materials, etc.), irrespective of their geometry and constitutions.

3. SECOND LAW OF THERMODYNAMICS

The second law of thermodynamics (the law of entropy) is an inequality which in its global form has the same structure as (2.4) except that the equality sign (=) in (2.4) is replaced by the inequality sign ($\geq$). In this case we have the identification[1]

$$\phi = \rho\eta \ , \quad \tau^k = q^k/\theta \quad , \ g = \rho h/\theta$$

where η and θ are, respectively, the entropy density per unit mass, and the absolute temperature subject to

$$\theta > 0 \quad , \qquad \inf\theta = 0$$

The localization now gives

$$\rho\dot{\eta} - \left(\frac{q^k}{\theta}\right)_{,k} - \frac{\rho h}{\theta} \geq \rho\hat{b} - \hat{\rho}\eta$$

$$[\rho\eta(v^k - u^k) - \frac{q^k}{\theta} - \hat{B}^k]n_k \geq 0 \tag{3.1}$$

$$\int_{V-\sigma} \rho\hat{b}\, dv + \int_{\sigma} [\hat{B}^k]\, da_k = 0$$

where $\hat{b}$ and $\hat{B}^k$ are the volume and surface localization residuals.

We now transform (3.1) into a more convenient form by introducing the Helmholtz free energy function

$$\psi = \varepsilon - \theta\eta \tag{3.2}$$

Elimination of h between $(2.12)_1$ and $(3.1)_1$ gives

$$- \frac{\rho}{\theta}(\dot{\psi} + \dot{\theta}\eta) + \frac{1}{\theta} t^k \cdot v_{,k} + \frac{1}{\theta^2} q^k \theta_{,k} - \frac{\hat{\rho}}{\theta}(\psi - \frac{1}{2} v \cdot v) \tag{3.3}$$
$$- \frac{\rho}{\theta} \hat{f} \cdot v + \rho\left(\frac{\hat{h}}{\theta} - \hat{b}\right) \geq 0 \qquad \text{in } V - \sigma$$

*This identification for τ^k is not necessary. In classical elasticity, the entropy inequality leads again to $\tau^k = q^k/\theta$.

The inequality (3.3) is fundamental for nonlocal thermodynamic processes. It is the generalization of the celebrated inequality of Clausius-Duhem. We shall use this inequality to arrive at the specific forms of the constitutive equations. *The entropy inequality* (3.3) *is postulated to be valid for all independent processes.*

4. CONSTITUTIVE EQUATIONS

The balance laws formulated in Section 2 are inadequate for the determination of motions of nonlocal elastic solids. The nature of the media must be characterized by means of a set of constitutive equations. We need to construct equations for the *constitutive dependent variables* $t^{k\ell}$, q^k, ψ and η . In addition, the nonlocal residuals must be determined. Disregarding the residuals on the discontinuity surface σ we have $\hat{\rho}$, $\hat{f}$ and $\hat{h}$ to be determined. Since the nonlocal fields at all points of the body affect all these variables, we must generalize the classical constitutive equations to include the effects of fields at all points. Thus we define nonlocal elastic solids by constitutive equations of the form

$$\psi = \psi(\underset{\sim}{x}', \underset{\sim}{x}'_{,K}, \theta') \tag{4.1}$$

where a prime abbreviates the *functional being over all material points* $\underset{\sim}{X}'$ *of the body*, i.e.,

$$\underset{\sim}{x}' \equiv \underset{\sim}{x}(\underset{\sim}{X}', t), \quad \underset{\sim}{x}'_{,K} \equiv \underset{\sim}{x}_{,K'}(\underset{\sim}{X}', t), \quad \theta' \equiv \theta(\underset{\sim}{X}', t)$$

When the local variables are needed, we drop the prime and write

$$\underset{\sim}{x} \equiv \underset{\sim}{x}(\underset{\sim}{X}, t) \quad , \quad \underset{\sim}{x}_{,K} \equiv \underset{\sim}{x}_{,K}(\underset{\sim}{X}, t), \theta \equiv \theta(\underset{\sim}{X}, t)$$

Equations of the form (4.1) are also written for $t^{k\ell}$, q^k, and η.

We distinguish the nonlocal residuals from the constitutive response functionals in that they may not depend on some of these variables. In fact we postulate (Eringen [1974a]):

Axiom (i) (*Nonlocality*). *The nonlocal residuals cannot depend on the state of the system at* $\underset{\sim}{X}$, *explicitly.* Since these residuals arise as a result of the influence of *all* material points, $\underset{\sim}{X}'$, of the body to the point $\underset{\sim}{X}$, we expect that they can depend only on the difference values of the state variables at $\underset{\sim}{X}'$ and $\underset{\sim}{X}$. For example, in Newtonian mechanics the interatomic force is a function of the distance $|\underset{\sim}{x}'-\underset{\sim}{x}|$, between two points $\underset{\sim}{x}'$ and $\underset{\sim}{x}$ and not a function of $\underset{\sim}{x}$ alone. The body force $\underset{\sim}{f}$ at $\underset{\sim}{x}$ has already been counted in the equation of motion. Thus, $\hat{\underset{\sim}{f}}$ cannot depend on the state variables θ and C_{KL} at $\underset{\sim}{X}$.

A second axiom essential to the present treatment is the axiom of objectivity.

Axiom (ii) (Objectivity or Frame Indifference). The constitutive functionals and the nonlocal residuals are invariant under arbitrary time-dependent motions of the spatial frame of reference and constant shift of time:

$$\bar{\mathbf{x}}(\mathbf{X},\bar{t}) = \mathbf{Q}(t)\mathbf{x}(\mathbf{X},t) + \mathbf{b}(t)$$
$$\mathbf{Q}\mathbf{Q}^T = \mathbf{Q}^T\mathbf{Q} = \mathbf{I}, \quad \det \mathbf{Q} = 1 \quad , \bar{t} = t-a \tag{4.2}$$

Since the nonlocal residuals are the result of the long-range interatomic effects, it is expected that these residuals do not change when the entire body undergoes arbitrary rigid motions.

From the axiom of objectivity, it follows that

$$\psi(\mathbf{Q}\mathbf{x} + \mathbf{b}, \mathbf{Q}\mathbf{x}'_{,K}, \theta) = \psi(\mathbf{x}', \mathbf{x}'_{,K}, \theta')$$

For arbitrary vectors $\mathbf{b}$, this implies that ψ must be a functional of *not* $\mathbf{x}'$ but $\boldsymbol{\chi}' \equiv \mathbf{x}'-\mathbf{x}$. For arbitrary $\mathbf{Q}$ satisfying $(4.2)_{2,3}$ it also implies that ψ must be a functional in the scalar products

$$\boldsymbol{\chi}' \cdot \boldsymbol{\chi}', \quad C'_K \equiv \boldsymbol{\chi}' \cdot \mathbf{x}'_{,K} , \quad C'_{KL} \equiv \mathbf{x}'_{,K} \cdot \mathbf{x}'_{,L} \tag{4.3}$$

The first of these can be expressed in terms of $\mathbf{C}'_K$ and $\mathbf{C}'_{KL}$, for

$$\boldsymbol{\chi}' \cdot \boldsymbol{\chi}' = (C'_K X'^K_{,k})(C'_L X'^L_{,k}) = C'_K C'_L \overset{-1}{C}{}'^{KL}$$

Thus we have

$$\psi = \psi(C'_K, C'_{KL}, \theta') \tag{4.4}$$

To separate the local effects we may equivalently write

$$\psi = \psi(C'_K, \tilde{C}{}'^{X}_{KL}, \overset{\sim}{\theta}{}'^{X}; C_{KL}, \theta) \tag{4.5}$$

where

$$\tilde{C}{}'^{X}_{KL} \equiv C'_{KL} - C_{KL} , \quad \theta'^{X} \equiv \theta' - \theta, \quad \tilde{C}{}'^{X}_{K} \equiv C'^{X}_K - C_K = C'_K \tag{4.6}$$

are the *difference nonlocality measures*. Thus the constitutive equations of nonlocal elastic solids obeying the axiom of objectivity must be of the form (4.5) where ψ is a functional on the functions carrying a prime and a function of unprimed variables.

We write (4.5) in brief form

$$\psi = \psi(F'; G) \tag{4.7}$$

where

$$F' \equiv \{C'_K, C'^{X}_{KL}, \theta'^{X}\} \quad , \; G \equiv \{C_{KL}, \theta\} \tag{4.8}$$

and assume that these functions are continuously differentiable with respect to their arguments. Next we assume that the space of functions F' is a linear normed space. In fact it is an *inner product (Hilbert) space, H,* in which the inner product of two such functions F'_1 and F'_2 is defined by

$$(F'_1, F'_2)_H = \int_{V-\Sigma} H(|\underset{\sim}{X}'-\underset{\sim}{X}|)F'_1(\underset{\sim}{X}')F'_2(\underset{\sim}{X}')dV(\underset{\sim}{X}') \tag{4.9}$$

where $H(|\underset{\sim}{X}'-\underset{\sim}{X}|)$ is a *positive decreasing function* of $|\underset{\sim}{X}'-\underset{\sim}{X}|$ such that $H(0) = 1$. The *influence* function H so introduced helps to emphasize the influence of motions of material points near to $\underset{\sim}{X}$. This is in accordance with the nature of intermolecular forces since they die out with the distance from the point $\underset{\sim}{X}$ at which ψ is to be evaluated. Among various possible forms for H we cite

$$H(|\underset{\sim}{X}'-\underset{\sim}{X}|) = \exp\{-\alpha(\underset{\sim}{X})|\underset{\sim}{X}'-\underset{\sim}{X}|\}, \; \alpha(\underset{\sim}{X}) > 0 \tag{4.10}$$

Other possibilities were discussed in our previous works (cf. Eringen [1966], [1974]).

For a functional $\psi(F')$, the Fréchet derivative $\delta\psi(F'|\Omega')$ is defined by

$$\lim_{||\Omega'||_H \to 0} |\psi(F'+\Omega')-\psi(F')-\delta\psi(F'|\Omega')|=0, F',\Omega' \in H \tag{4.11}$$

where $||\Omega'||_H$ represent the norm defined by

$$||\Omega'||_H = (\Omega', \Omega')^{1/2}_H \tag{4.12}$$

The Fréchet derivative is a linear functional in Ω'. Since F' and Ω' belongs to H, according to a theorem in the Hilbert space theory, we have the representation of the form

$$\delta\psi(F'|\Omega') = \int_{V-\Sigma} \frac{\delta\psi}{\delta F'}(F';\underset{\sim}{\Lambda})\cdot\Omega'(\underset{\sim}{\Lambda})dV(\underset{\sim}{\Lambda}) \tag{4.13}$$

where $\delta\psi/\delta F'$ is a functional of F' and a function of a vector $\underset{\sim}{\Lambda}$ and the integration is over the volume V of B in the reference state excluding the discontinuity surface Σ(the image of σ). Henceforth we also write $dV(\underset{\sim}{\Lambda}) \equiv dV'$, for brevity.

With this apparatus at hand, and assuming that ψ is also continuously differentiable with respect to G, we can calculate

$$\dot{\psi} = \frac{\partial\psi}{\partial G} \cdot \dot{G} + \int_{V-\Sigma} \frac{\delta\psi}{\delta F'} (F';G,\underset{\sim}{\Lambda}) \cdot \dot{F}'(\underset{\sim}{\Lambda})dV' \tag{4.14}$$

Substituting this into the entropy inequality (3.3) and using (4.8) we will have

$$- \frac{\rho}{\theta} (\frac{\partial\psi}{\partial\theta} - \int_{V-\Sigma} \frac{\delta\psi}{\delta\overset{'X}{\underset{\sim}{\theta}}} dV' + \eta)\dot{\theta} + \frac{1}{\theta} [t^{k\ell} - 2\rho(\frac{\partial\psi}{\partial C_{KL}}$$

$$- \int_{V-\Sigma} \frac{\delta\psi}{\delta \overset{'X}{\underset{\sim}{C}}_{KL}} dV')x^k_{,K}x^\ell_{,L}] v_{\ell,k} - \frac{\hat{\rho}}{\theta} (\psi - \frac{1}{2} \underset{\sim}{v}\cdot\underset{\sim}{v})$$

$$- \frac{\rho}{\theta} (\hat{f}^k - \int_{V-\Sigma} \frac{\delta\psi}{\delta \overset{'}{C}_K} \overset{'k}{x}_{,K} dV')v_k + \frac{1}{\theta^2} q^k\theta_{,k}$$

$$- \frac{\rho}{\theta} \int_{V-\Sigma} \{ \frac{\delta\psi}{\delta\overset{'X}{\underset{\sim}{\theta}}} \dot{\theta}' + \frac{\delta\psi}{\delta\overset{'X}{\underset{\sim}{C}}_{KL}} \dot{C}'_{KL} + \frac{\delta\psi}{\delta \overset{'}{C}_K} [v'_k + (x'_\ell - x_\ell)v'^\ell_{,k}]\overset{'k}{x}_{,K}\} dV'$$

$$+ \frac{\hat{\rho}}{\theta} - \rho\hat{b} \geq 0 \tag{4.15}$$

This inequality is linear in $\dot{\theta}$, $v_{\ell,k}$, $\theta_{,k}$ and quadratic in v_k since all constitutive functionals are independent of these and the residuals cannot depend on these *local* variables.

The necessary and sufficient conditions for the inequality not to be violated for all possible variations of these functions are

$$\eta = - \frac{\partial\psi}{\partial\theta} + \int_{V-\Sigma} \frac{\delta\psi}{\delta\overset{'X}{\underset{\sim}{\theta}}} dV'$$

$$t^{k\ell} = 2\rho(\frac{\partial\psi}{\partial C_{KL}} - \int_{V-\Sigma} \frac{\delta\psi}{\delta\overset{'X}{\underset{\sim}{C}}_{KL}} dV')x^k_{,K}x^\ell_{,L}$$

$$q^k = 0 \tag{4.16}$$

$$\hat{f}^k = \int_{V-\Sigma} \frac{\delta\psi}{\delta\overset{'X}{\underset{\sim}{C}}_K} \overset{'k}{x}_{,K} dV'$$

$$\hat{\rho} = 0^1$$

and

---

[1] Actually from (4.15) it follows that $\hat{\rho}/\theta \geq 0$. However since $\theta>0$ and $\int_{V-\sigma} \hat{\rho}\, dv = 0$ (see $(2.8)_3$ in the absence of $\hat{R}^k$ or σ) there follows $\hat{\rho}=0$.

$$\frac{\rho}{\theta}(\hat{h} - \hat{D}) - \rho\hat{b} \geq 0 \qquad (4.17)$$

where

$$\hat{D} \equiv \int_{V-\Sigma} \{\frac{\delta\psi}{\delta \overset{'X}{\underset{\sim}{\theta}}} \dot{\theta}' + \frac{\delta\psi}{\delta \overset{'X}{\underset{\sim}{C}}_{KL}} \dot{C}'_{KL} + \frac{\delta\psi}{\delta \overset{'K}{C}} [v'_k + (x'_\ell - x_\ell)\overset{'\ell}{v}_{,k}]\overset{'k}{x}_{,K}\} \, dV' \qquad (4.18)$$

is the *nonlocal dissipation function.*

According to the axiom of objectivity, $\hat{h}$ and $\hat{b}$ must be invariant under rigid motions of the body. If the entropy inequality (4.17) is to remain invariant under such motions then it follows that $\hat{D}$ must also be invariant under rigid motions. Hence

$$\frac{\delta\psi}{\delta C'_K} = 0 \qquad (4.19)$$

and therefore ψ cannot depend on C'_K and the nonlocal body force vanishes, i.e.,

$$\psi = \psi(\overset{'X}{\underset{\sim}{C}}_{KL}, \overset{'X}{\underset{\sim}{\theta}}; C_{KL}, \theta) \quad , \; \hat{\underset{\sim}{f}} = \underset{\sim}{0} \qquad (4.20)$$

We have therefore proved

Theorem. The constitutive functionals and the nonlocal residuals of the nonlocal elasticity do not violate the second law of thermodynamics and the axiom of objectivity if and only if they are of the form (4.16) *and* (4.20) *subject to* (4.17) *and* (4.19).

Upon integrating (4.17) over $V-\sigma$ we will have

$$\int_{V-\sigma} \frac{\rho}{\theta}(\hat{h} - \hat{D})dv = 0 \qquad (4.21)$$

This represents the total dissipation of the elastic body. If the elastic body is to be dissipation free (as we expect), then this integral must vanish. A sufficient condition for this is[1]

$$\hat{h} = \hat{D} \qquad (4.22)$$

5. LINEAR THEORY OF NONLOCALITY

To construct the constitutive equations of the linear theory of nonlocal elastic solids we take

[1]While this is not necessary, it can be shown that this result is necessary and sufficient for the linear theory (cf. Eringen [1974a].

$$\rho_o \psi \equiv \Sigma = \Sigma_o(\underset{\sim}{C},\theta,\underset{\sim}{X}) + \int_{V-\Sigma} [\Sigma'_o(\underset{\sim}{C},\theta,\underset{\sim}{X},\underset{\sim}{\Lambda})\theta^{'X}(\underset{\sim}{\Lambda}) + \Sigma_2^{'KL}(\underset{\sim}{C},\theta,\underset{\sim}{X},\underset{\sim}{\Lambda})C_{KL}^{'X}(\underset{\sim}{\Lambda})]dV(\underset{\sim}{\Lambda}) \tag{5.1}$$

Substituting this into (4.16) we get

$$\eta = -\frac{1}{\rho_o}\frac{\partial\Sigma_o}{\partial\theta} - \frac{1}{\rho_o}\int_{V-\Sigma}\left[\frac{\partial\Sigma'_o}{\partial\theta}(\theta'-\theta) + \frac{\partial\Sigma_2^{'KL}}{\partial\theta}(C'_{KL}-C_{KL}) - \Sigma'_o\right]dV'$$

$$T^{KL} \equiv \frac{\rho_o}{\rho} t^{k\ell} X^K_{,k} X^L_{,\ell} = 2\frac{\partial\Sigma_o}{\partial C_{KL}} + 2\int_{V-\Sigma}\left[\frac{\partial\Sigma'_o}{\partial C_{KL}}(\theta'-\theta) + \frac{\partial\Sigma_2^{'MN}}{\partial C_{KL}}(C'_{MN}-C_{MN}) - \Sigma_2^{'KL}\right]dV' \tag{5.2}$$

To obtain the *linear theory* we further set

$$\begin{aligned}
\Sigma_o &= \Sigma_{oo} + \Sigma_{o1}^{KL}E_{KL} + \frac{1}{2}\Sigma_{o2}^{KLMN}E_{KL}E_{MN} \\
\Sigma'_o &= \Sigma'_{oo} + \Sigma_{o1}^{'KL}E_{KL} \\
\Sigma_2^{'KL} &= \Sigma_{2o}^{'KL} + \frac{1}{2}\Sigma_{21}^{'KLMN}E_{MN}
\end{aligned} \tag{5.3}$$

where Σ_{oo}, Σ_{o1}^{KL}, Σ_{o2}^{KLMN} are functions of θ and $\underset{\sim}{X}$ and Σ'_{oo}, $\Sigma_{o1}^{'KL}$, $\Sigma_{2o}^{'KL}$ and $\Sigma_{21}^{'KLMN}$ are functions of $\theta, \underset{\sim}{X}$ and $\underset{\sim}{\Lambda}$. In addition we have the obvious symmetry relations

$$\begin{aligned}
&\Sigma_{o1}^{KL} = \Sigma_{o1}^{LK}, \quad \Sigma_{o2}^{KLMN} = \Sigma_{o2}^{LKMN} = \Sigma_{o2}^{KLMN} = \Sigma_{o2}^{MNKL} \\
&\Sigma_{o1}^{'KL} = \Sigma_{o1}^{'LK}, \quad \Sigma_{2o}^{'KL} = \Sigma_{2o}^{'LK} \\
&\Sigma_{21}^{'KLMN} = \Sigma_{21}^{'LKMN} = \Sigma_{21}^{'KLMN} = \Sigma_{21}^{'MNKL}
\end{aligned} \tag{5.4}$$

In the spirit of the linear theory we also set

$$C_{KL} \simeq \delta_{KL} + 2E_{KL} = \delta_{KL} + U_{K,L} + U_{L,K} \tag{5.5}$$

where U_K is the *displacement vector*. Introducing (5.3) and (5.5) into (5.2) we obtain

$$\eta = -\frac{1}{\rho_o}\frac{\partial\Sigma_{oo}}{\partial\theta} - \frac{1}{\rho_o}\frac{\partial\Sigma^{KL}_{o1}}{\partial\theta}E_{KL} - \frac{1}{\rho_o}\int_{V-\Sigma}\Big[\frac{\partial\Sigma'_{oo}}{\partial\theta}(\theta'-\theta) + \frac{\partial\Sigma'^{KL}_{2o}}{\partial\theta}(E'_{KL} - E_{KL}) - \Sigma'_{oo} - \Sigma'^{KL}_{o1}E_{KL}\Big]dV'$$

$$t^{k\ell} = \frac{\rho}{\rho_o}\{\Sigma^{KL}_{o1} + \Sigma^{KLMN}_{o2}E_{MN} + \int_{V-\Sigma}[\Sigma'^{KL}_{o1}(\theta'-\theta) + \Sigma'^{KLMN}_{21}(E'_{MN}-E_{MN}) - 2E'^{KL}_{2o} - \Sigma'^{KLMN}_{21}E_{MN}]dV'\}x^k_{,K}x^\ell_{,L} \quad (5.6)$$

In the linear theory we also have

$$\rho/\rho_o \simeq 1 - e^k_{\ k},\quad e_{k\ell} \simeq \frac{1}{2}(u_{k,\ell}+u_{\ell,k}),\quad r_{k\ell} \simeq \frac{1}{2}(u_{k,\ell}-u_{\ell,k})$$

$$x^k_{,K} \simeq (\delta_{MK}+U_{M,K})\delta^M_{\ k},\quad X^K_{,k} \simeq (\delta_{MK}-U_{M,K})\delta^M_{\ k} \quad (5.7)$$

$$E_{KL} \simeq e_{k\ell}x^k_{,K}x^\ell_{,L},\quad u^k \simeq U^K\delta^k_{\ K}$$

where $\delta^k_{\ K}$, $\delta^K_{\ k}$ and δ_{MK} are the Kronecker deltas and $e_{k\ell}$ and $r_{k\ell}$ are, respectively, the eulerian strain and rotation tensors expressed in terms of the components of eulerian displacement vectors, u^k. On using (5.7) in $(5.6)_2$ we will have

$$t^{k\ell} = \alpha^{k\ell}(1-e^r_{\ r})+\alpha^{km}(e^\ell_{\ m}+r^\ell_{\ m})+\alpha^{\ell m}(e^k_{\ m}+r^k_{\ m}) + \sigma^{k\ell mn}e_{mn} + \int_{V-\sigma}[\sigma'^{k\ell}(\theta'-\theta)+\sigma'^{k\ell mn}e'_{mn}]dv' \quad (5.8)$$

where the quantities carrying a prime are considered to be a function of a vector λ (replacing Λ)[1], $dv' \equiv dv(\lambda)$ and

---

[1] Note that in the transformation of the volume integral $dV' = (\rho'/\rho'_o)dv' \simeq (1-e'_{kk})dv'$. But e'_{kk} does not contribute to the integral in (5.8) since we drop the nonlinear terms.

$$
\begin{aligned}
\alpha^{k\ell} &\equiv (\Sigma_{o1}^{KL} - \int_{V-\Sigma} 2\Sigma_{2o}^{'KLMN} dV')\delta^k{}_K \delta^\ell{}_L \\
\sigma^{k\ell mn} &\equiv (\Sigma_{o2}^{KLMN} - 2\int_{V-\Sigma} \Sigma_{21}^{'KLMN} dV')\delta^k{}_K \delta^\ell{}_L \delta^m{}_M \delta^n{}_N \\
\sigma^{'k\ell} &\equiv \Sigma_{o1}^{'KL}\delta^k{}_K \delta^\ell{}_L \\
\sigma^{'k\ell mn} &\equiv \Sigma_{21}^{'KLMN}\delta^k{}_K \delta^\ell{}_L \delta^m{}_M \delta^n{}_N,
\end{aligned}
\tag{5.9}
$$

If the temperature is kept uniform throughout the body, then (5.8) reduces to

$$
\begin{aligned}
t^{k\ell} &= \alpha^{k\ell}(1-e^r{}_r)+\alpha^{km}(e^\ell{}_m+r^\ell{}_m)+\alpha^{\ell m}(e^k{}_m+r^k{}_m) \\
&\quad + \sigma^{k\ell mn}e_{mn} + \int_{V-\sigma} \sigma^{'k\ell mn'} e_{mn}\, dv'
\end{aligned}
\tag{5.10}
$$

This is the final form of the stress constitutive equations for *isothermal, anisotropic, inhomogeneous, linear, nonlocal elastic solids*.

We note the obvious symmetry conditions for the material moduli $\alpha^{k\ell}(\underset{\sim}{x})$, $\delta^{k\ell mn}(\underset{\sim}{x})$ and the nonlocal moduli $\sigma^{'k\ell mn}(\underset{\sim}{x},\underset{\sim}{\lambda})$:

$$
\begin{aligned}
&\alpha^{k\ell} = \alpha^{\ell k}, \quad \sigma^{k\ell mn} = \sigma^{\ell kmn} = \sigma^{k\ell nm} = \sigma^{mnk\ell} \\
&\sigma^{'k\ell mn} = \sigma^{'\ell kmn} = \sigma^{'k\ell nm} = \sigma^{'mnk\ell}
\end{aligned}
\tag{5.11}
$$

The following observations are significant to note:

(a) Even for non heat-conducting solids (i.e., $\underset{\sim}{q} = \underset{\sim}{0}$), temperature differences between distant points and $\underset{\sim}{x}$ produce stress at $\underset{\sim}{x}$ (cf. eq. (5.8)). This is as expected.

(b) Local and nonlocal symmetry conditions for stress may be different. For example, a material may be anisotropic in the atomic and molecular range while isotropic in the macroscopic scale.

(c) For fully isotropic solids (locally and nonlocally), (5.10) reduces to

$$
\begin{aligned}
t_{k\ell} &= \alpha(1-e^r{}_r)\delta_{k\ell} + \lambda e_{rr}\delta_{k\ell} + 2(\mu+\alpha)e_{k\ell} \\
&\quad + \int_{V-\sigma} (\lambda' e^{'r}{}_r \delta_{k\ell} + 2\mu' e'_{k\ell})dv'
\end{aligned}
\tag{5.12}
$$

since in this case we write

$$\alpha^{k\ell} = \alpha\delta^{k\ell} \ , \ \sigma^{k\ell mn} = \lambda\delta^{k\ell}\delta^{mn} + \mu(\delta^{km}\delta^{\ell n} + \delta^{kn}\delta^{\ell m})$$
$$\sigma'^{k\ell mn} = \lambda'\delta^{k\ell}\delta^{mn} + \mu'(\delta^{km}\delta^{\ell n} + \delta^{kn}\delta^{\ell n}) \tag{5.13}$$

where α, λ and μ are functions of $\underset{\sim}{x}$ only and λ' and μ' are functions of $\underset{\sim}{x}$ and a vector $\underset{\sim}{\lambda}$.

If the natural state is stress free, then $\alpha = 0$. If the material is homogeneous, then α, λ and μ are constants and λ' and μ' depend on $\underset{\sim}{x}$ only through $|x_k - \lambda_k|$, i.e.,

$$\lambda' = \lambda'(|x_k - \lambda_k|) \quad , \ \mu' = \mu'(|x_k - \lambda_k|) \tag{5.14}$$

These moduli must decrease fast as $|\underset{\sim}{x} - \underset{\sim}{\lambda}|$ becomes large (axiom of attenuating neighborhood). Several forms can be suggested, e.g.,

$$\lambda' = \lambda_o \exp(-k|\underset{\sim}{x} - \underset{\sim}{\lambda}|) \quad , \text{ real } k > 0 \tag{5.15}$$

Alternatively, as we shall see, the atomic theory of lattices or experiments on dispersion relations can be used to obtain the form of these moduli (cf. Eringen [1972c]).

6. FIELD EQUATIONS

We now derive the field equations of the linear, homogeneous nonlocal isothermal elastic solids. For this we only need to substitute (5.10) into $(2.9)_1$. Recalling (2.10) and the fact that $\hat{\rho} = 0$, $\hat{\underset{\sim}{f}} = \underset{\sim}{0}$,

$$\alpha_{km}u_{\ell,mk} + \sigma_{k\ell mn}u_{m,nk} + \int_{V-\sigma} \frac{\partial\sigma'_{k\ell mn}}{\partial x^k} e'_{mn}\, dv'$$
$$+ \rho(f_\ell - \ddot{u}_\ell) = 0 \tag{6.1}$$

Since the nonlocal moduli $\sigma'_{k\ell mn}$ is a function of $|\underset{\sim}{x} - \underset{\sim}{\lambda}|_k$, we may transform the volume integral by writing $\partial/\partial x^k = -\partial/\partial\lambda^k$ and use by-part integration. The generalized Green-Gauss theorem now gives

$$\alpha_{kn}u_{\ell,mk} + \sigma_{k\ell mn}u_{m,nk} + \int_{V-\sigma} \sigma'_{k\ell mn}u_{m,n'k'}\, dv' \tag{6.2}$$
$$- \oint_{S-\sigma} \tau'_{k\ell}da'_k + \int_\sigma [\tau'_{k\ell}]da'_k + \rho(f_\ell - \ddot{u}_\ell) = 0$$

where

$$\tau'_{k\ell} \equiv \sigma'_{k\ell mn} e'_{mn}$$

is the *surface stress tensor*.

For isotropic solids having a stress free natural state, (6.2) reduces to

$$(\lambda+\mu)u_{k,k\ell} + \mu u_{\ell,kk} + \int_{V-\sigma} [(\lambda'+\mu')u'_{k,k\ell} + \mu' u'_{\ell,kk}]dv' \tag{6.3}$$
$$- \oint_{S-\sigma} \tau'_{k\ell} da'_k + \int_\sigma [\tau'_{k\ell}] da'_k + \rho(f_\ell - \ddot{u}_\ell) = 0$$

where

$$\tau'_{k\ell} \equiv \lambda' e'_{rr} \delta_{k\ell} + 2\mu' e'_{k\ell} \tag{6.4}$$

These equations were obtained previously in a different way, cf. Eringen [1972c].

It is important to note the appearance of surface integrals in (6.2) and (6.3). These bring into play the *surface strains*. Thus the nonlocal theory inherently involves the *surface physics* so important in the discussion of many interesting phenomena that take place near the surface of a solid or in the interface of two solids. For example, the surface tension is naturally contained in the nonlocal theory, as discussed more specifically in a previous publication (Eringen [1972d]).

For some purposes, it may be convenient to write (5.12) and (6.3) in the forms

$$t_{k\ell} = \int_{V-\sigma} (\lambda' e'_{rr} \delta_{k\ell} + 2\mu' e'_{k\ell}) dv' \tag{6.5}$$

$$\int_{V-\sigma} [(\lambda'+\mu')u'_{k,k\ell} + \mu' u'_{\ell,kk}]dv' - \oint_{S-\sigma} \tau'_{k\ell} da'_k \tag{6.6}$$
$$+ \int_\sigma [\tau'_{k\ell}] da'_k + \rho(f_\ell - \ddot{u}_\ell) = 0$$

with new interpretations of λ' and μ' which include local constants λ and μ.

7. PROPAGATION OF PLANE WAVES

Consider an elastic solid of infinite extent with no discontinuity surface or body forces. The field equations (6.6), by the usual Helmholtz decomposition,

$$\underset{\sim}{u} = \nabla\phi + \nabla \times \underset{\sim}{\psi} \ , \ \nabla \cdot \underset{\sim}{\psi} = 0 \tag{7.1}$$

is satisfied if the scalar and vector potentials ϕ and $\underset{\sim}{\psi}$ satisfy

$$\int_{V-\sigma} \ell'(|\underset{\sim}{\xi}-\underset{\sim}{x}|)\nabla'^2\phi' dv(\underset{\sim}{\xi}) = c_1^{-2}\ddot{\phi} \ , \tag{7.2}$$

$$\int_{V-\sigma} m'(|\underset{\sim}{\xi}-\underset{\sim}{x}|)\nabla' \times \nabla' \times \underset{\sim}{\psi}' \, dv(\underset{\sim}{\xi}) = c_2^{-2}\ddot{\underset{\sim}{\psi}} \tag{7.3}$$

where

$$\ell' \equiv \frac{\lambda'+2\mu'}{\lambda+2\mu} \ , \qquad m' \equiv \frac{\mu'}{\mu}$$
$$c_1^2 = \frac{\lambda+2\mu}{\rho} \ , \qquad c_2^2 \equiv \frac{\mu}{\rho} \tag{7.4}$$

where ∇' represents the gradient operator with respect to ξ. We now seek the plane wave solution of (7.2). To this end we introduce the Fourier transform

$$\overline{\phi}(\underset{\sim}{\xi}) = \int \int_{-\infty}^{\infty} \int \phi(x)\exp(i\underset{\sim}{\xi}\cdot\underset{\sim}{x}+i\omega t)d\xi_1 d\xi_2 d\xi_3$$

Applying the Fourier transform to (7.2) and (7.3) we obtain

$$\omega^2/c_1^2 = \overline{\ell}'(\underset{\sim}{\xi})(\xi_1^2 + \xi_2^2 + \xi_3^2) \tag{7.5}$$

$$\omega/c_2^2 = \overline{m}'(\underset{\sim}{\xi})(\xi_1^2 + \xi_2^2 + \xi_3^2) \tag{7.6}$$

where $\overline{\ell}'$ and $\overline{m}'$ are respectively the Fourier transforms of $\ell'(\underset{\sim}{x})$ and $m'(\underset{\sim}{x})$.

Equations (7.5) and (7.6) are the *dispersion relations* for the plane longitudinal and transverse waves. For the one dimensional case, (7.5) reads

$$\omega^2/c_1^2 = \overline{\ell}_1'(\xi)\xi^2 \tag{7.7}$$

where now $\overline{\ell}_1'(\xi)$ is defined by

$$\overline{\ell}_1'(\xi) = \int_{-\infty}^{\infty} \ell_1'(x)\exp(i\xi x)dx \tag{7.8}$$

in which

$$\ell_1'(|x_1-\xi_1|) \equiv \iint_{-\infty}^{\infty} \ell'(|\underset{\sim}{x}-\underset{\sim}{\xi}|)d\xi_2 d\xi_3 \tag{7.9}$$

The dispersion relations (7.7) may be compared with those known in lattice dynamics to obtain the form of the nonlocal moduli ℓ' and m'. Eringen [1972c] has already given some calculations when the form (6.3) is used. Here we repeat a similar calculation for the form (6.6) which leads to a simpler kernel function $\ell'(x)$.

From the atomic theory of a lattice for a nearest neighborhood interaction we have (cf. Brillouin [1946, p. 31])

$$\omega^2 = \frac{4U''}{M} \sin^2 \frac{\xi d}{2} \quad , \quad -\frac{\pi}{d} \leq \xi \leq \frac{\pi}{d}$$

where M is the mass of an atom in a one dimensional chain of atoms having equal distance d, attached to each other by linear springs whose spring constant is U". The phase velocity for a one dimensional local continuum is given by

$$c_1 = d\sqrt{U''/M} \tag{7.10}$$

so that

$$\frac{\omega^2}{c_1^2} = \frac{4}{d^2} \sin^2 \frac{\xi d}{2} \quad , \quad -\frac{\pi}{d} \leq \xi \leq \frac{\pi}{d} \tag{7.11}$$

The graph of ω within one Brillouin zone is shown in Fig. 1. Equating this to (7.7) we obtain

$$\bar{\ell}_1'(\xi) = \frac{4}{\xi^2 d^2} \sin^2 \frac{\xi d}{2}$$

whose inverse transform gives

$$\begin{aligned} \ell_1'(x) &= \frac{1}{d}\left(1 - \frac{|x|}{d}\right) \quad \text{when } \frac{|x|}{d} < 1 \\ &= 0 \qquad\qquad\qquad \text{when } \frac{|x|}{d} > 1 \end{aligned} \tag{7.12}$$

The graph of $\ell_1'(x)d$ is shown in Fig. 2. We thus find that the nonlocal theory gives a dispersion curve *identical* to that of the atomic theory of lattices, if the nonlocal moduli $\ell_1'(x)$ is of the form (7.12). This is a very simple result and yet extremely significant in that it points out the applicability of the nonlocal theory in the *entire* Brillouin zone for all wave lengths up to a magnitude comparable with atomic distances. As is well known, the classical theory of elasticity predicts no dispersion and is valid only for the infinite wave lengths (near $\xi = 0$, in Fig. 1).

Clearly one need not use lattice dynamics to determine the nonlocal material moduli. Equally one may benefit from the results of the neutron diffusion experiments abundantly available

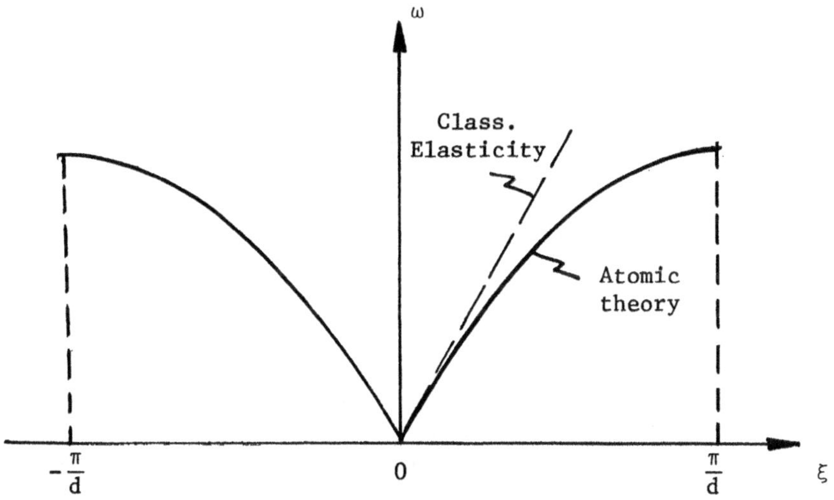

Fig. 1. Dispersion Curve in One Brillouin Zone.

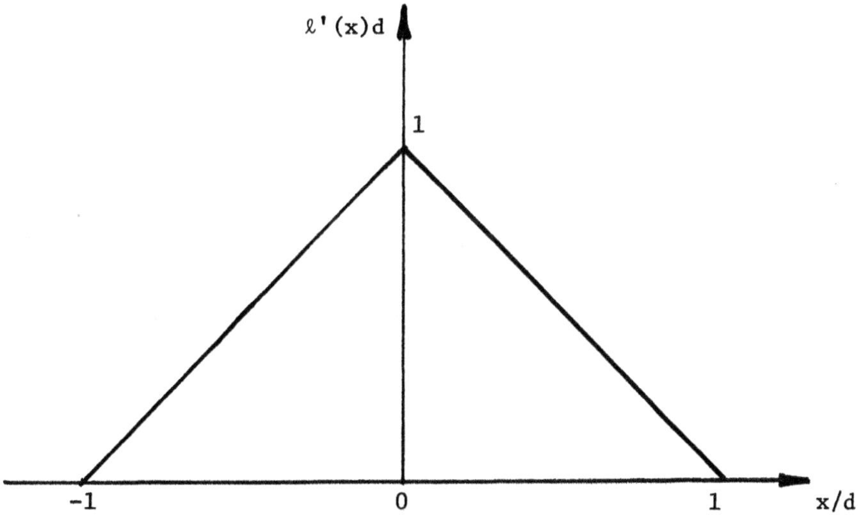

Fig. 2. Nonlocal Elastic Modulus.

in the literature for the dispersion of phonons. Moreover, for imperfect lattices the nonlocal theory provides the apparatus with its nonlocal moduli to be determined experimentally. Hence difficult problems involving imperfect lattices with dislocations and impurities can now be tackled mathematically which defied lattice dynamical calculations. Inherent in this scheme is also the capability of treatment of boundary conditions for lattices with finite boundaries. In fact surface characteristics of solids, hence the relevant surface mechanics, can now be handled by a direct approach.

Similarly for the vector waves one may employ (7.3) to obtain the nonlocal moduli m'(x) through a comparison of the dispersion curve with those available to us from the lattice dynamics. Since the method of approach is similar to the foregoing treatment, we do not pursue this matter further. Instead we give the expression of the free energy function for the one dimensional case treated above.

Using (5.1) for the form (6.5), we have

$$\rho_o\psi = \frac{1}{2} e_{k\ell} \int_V (\lambda' e'_{rr}\delta_{k\ell} + 2\mu' e'_{k\ell}) dv' = \frac{1}{2} t_{k\ell} e_{k\ell} \tag{7.13}$$

For the one dimensional problem this reads

$$\rho_o\psi = \frac{1}{2}\frac{\partial u}{\partial x}\int_{-\infty}^{\infty} (\lambda'+2\mu')\frac{\partial u}{\partial \xi}\, d\xi$$

$$= \frac{\lambda+2\mu}{2d^2}\frac{\partial^2\phi}{\partial x^2}\int_{x-d}^{x+d} \left(1 - \frac{|x-\xi|}{d}\right)\frac{\partial^2\phi}{\partial \xi^2}\, d\xi$$

Integration gives

$$\rho_o\psi = \frac{\lambda+2\mu}{2d^2}\frac{\partial^2\phi}{\partial x^2}[\phi(x+d) + \phi(x-d) - 2\phi(x)] \tag{7.14}$$

Employing Fourier representation of $\phi(x)$, this may be written as

$$\psi = -\frac{1}{4\pi}\frac{\partial^2\phi}{\partial x^2}\int_{-\pi/d}^{\pi/d} \bar{\phi}(\xi)\omega^2 \exp(-i\xi x - i\omega t)\, d\xi \tag{7.15}$$

or the total free energy becomes

$$\Psi \equiv \int_{-\infty}^{\infty} \psi dx = \frac{1}{4\pi}\int_{-\pi/d}^{\pi/d} \xi^2\omega^2 |\bar{\phi}(\xi)|^2\, d\xi \tag{7.16}$$

Thus the energy density per wave number, in one Brillouin zone, is

$$\psi_d(\xi) = \frac{1}{4\pi}\xi^2\omega^2(\xi)|\bar{\phi}(\xi)|^2 , \quad \frac{-\pi}{d} \le \xi \le \frac{\pi}{d} \tag{7.17}$$

For the classical elasticity, on the other hand, we have

$$\psi_o(\xi) = \frac{1}{4\pi}\,\xi^2\omega_o^2\,|\overline{\phi}(\xi)|^2 \quad , \quad -\infty \leq \xi \leq \infty \tag{7.18}$$

where

$$\omega_o = \left(\frac{\lambda+2\mu}{\rho}\right)^{1/2}\xi = c_1\xi \tag{7.19}$$

Thus

$$\frac{\psi_d(\xi)}{\psi_o(\xi)} = \frac{\omega^2(\xi)}{\omega_o^2(\xi)} \quad , \quad -\frac{\pi}{d} \leq \xi \leq \frac{\pi}{d} \tag{7.20}$$

We therefore see that the energy density per wave number is proportional to $\omega^2(\xi)$. In the limit, as the atomic distance $d \to 0$, we obtain the classical result, i.e.,

$$\lim_{d\to o} \psi_d(\xi) = \psi_o(\xi) \quad , \quad -\infty \leq \xi \leq \infty \tag{7.21}$$

8. SURFACE WAVES

According to the classical linear elasticity, the Rayleigh surface waves that propagate on the surface of an isotropic elastic half space are non dispersive. The Rayleigh surface wave velocity, for the Poisson material ($\lambda=\mu$), is, in fact, given by

$$c_R^o \simeq 0.9194\,\sqrt{\mu/\rho} \tag{8.1}$$

where λ and μ are Lamé constants and ρ is the mass density. Yet when atomic theory of lattices is used, it is found that the surface waves are dispersive, cf. Gazis et al. [1960]; Maradudin, et al. [1971, p. 531]. Surface waves being generally of high frequency, it is important to understand their nature in the region of small wave lengths. Irrespective of the importance of this matter in the solid state physics, it may be useful in dealing with the earthquake phenomena and underground explosions, since for small wave lengths the granular nature of earth must be considered. While lattice dynamic of the problem treatment exist, it is not very useful on account of the facts that: (i) lattice dynamic calculations are often extremely tedious, and (ii) the lattice dynamic calculations cannot be extended to other problems involving more complicated boundary conditions and/or complicated inner structure involving dissymmetrics and inhomogeneities.

The nonlocal theory provides an excellent resolution for both of these problems, especially in regard to (ii), by having a kernel function which can accommodate all sorts of complications in the structure of the body and yet as a continuum theory allows the treatment of complicated boundary and initial conditions. Here we give a brief discussion based on our recent work [1973d][1].

Consider an elastic half-space occupying the region $x_2 \geq 0$ of the rectangular frame of reference x_k $(k = 1,2,3)$. The problem requires finding a solution of the system

$$t_{k\ell,k} = \rho \ddot{u}_\ell \tag{8.2}$$

$$t_{k\ell} = \int_V [\lambda' u'_{r,r} \delta_{k\ell} + \mu' (u'_{k,\ell} + u'_{\ell,k})] dv' \tag{8.3}$$

under the boundary conditions

$$\begin{aligned} t_{22} = t_{21} &= 0 \qquad \text{when } x_2 = 0 \\ u_k &= 0 \qquad \text{as } \quad x_2 \to \infty \end{aligned} \tag{8.4}$$

For cylindrical disturbances with x_3-axis, the problem becomes two-dimensional in the region $-\infty < x_1 < \infty$, $0 \leq x_2 < \infty$. In this case, $u_k = u_k(x_1, x_2, t)$, $k = 1,2$ and $u_3 = 0$, and the volume integral in (8.3) reduces to a surface integral over x'_1 and x'_2.

We are interested in understanding small wave length behavior of surface waves in the x_1-direction. Since these waves are confined to a very thin layer near $x_2 = 0$, we assume a δ-function behavior for λ' and μ' in the direction x_2. Thus we take

$$\lambda' = \lambda(|x'_1 - x_1|)\delta(|x'_2 - x_2|), \quad \mu' = \mu(|x'_1 - x_1|)\delta(|x'_2 - x_2|) \tag{8.5}$$

With this (8.3) further reduces to

$$t_{k\ell} = \int_{-\infty}^{\infty} [\lambda(|x'_1 - x_1|) u'_{r,r} \delta_{k\ell} + \mu(|x'_1 - x_1|)(u'_{k,\ell} + u'_{\ell,k})]_{x'_2 = x_2} dx'_1 \tag{8.6}$$

By means of Fourier representation

$$u_k(x_1, x_2, t) = \frac{1}{2\pi} \iint_{-\infty}^{\infty} \bar{u}_k(\xi, x_2, \omega) \exp[-i(\xi x_1 + \omega t)] d\xi d\omega \tag{8.7}$$

(8.2) and (8.6) may be reduced to

---

*Here we employ the nonlocal modulus $\ell'_1(x)$ given by (7.12) which allows no curve fitting at all.

$$-i\xi\bar{t}_{11} + \bar{t}_{21,2} + \rho\omega^2\bar{u}_1 = 0$$

$$-i\xi\bar{t}_{12} + \bar{t}_{22,2} + \rho\omega^2\bar{u}_2 = 0$$

$$\bar{t}_{11} = -i\xi(\bar{\lambda}+2\bar{\mu})\bar{u}_1 + \bar{\lambda}\,\bar{u}_{2,2} \tag{8.8}$$

$$\bar{t}_{22} = -i\xi\bar{\lambda}\bar{u}_1 + (\bar{\lambda}+2\bar{\mu})\bar{u}_{2,2}$$

$$\bar{t}_{12} = \bar{\mu}(\bar{u}_{1,2} - i\xi\bar{u}_2)$$

We now seek a solution of this system in the form

$$\bar{u}_k(\xi,x_2,\omega) = \bar{U}_k(\xi,\omega)e^{-\alpha x_2} \quad , \; \alpha > 0 \tag{8.9}$$

Through (8.8) we obtain the general solution for $\bar{u}_k$

$$\begin{aligned} \bar{u}_1 &= \bar{U}_{11}\, e^{-\alpha_1 x_2} + \bar{U}_{12}\, e^{-\alpha_2 x_2} \\ \bar{u}_2 &= \bar{U}_{11}\, \gamma_1\, e^{-\alpha_1 x_2} + \bar{U}_{12}\gamma_2 e^{-\alpha_2 x_2} \end{aligned} \tag{8.10}$$

where $\bar{U}_{mn}(\xi,\omega)$ are constants of integrations,

$$\alpha_1^2 = \xi^2 - h^2 \quad , \quad \alpha_2^2 = \xi^2 - k^2$$

$$\gamma_j \equiv -\frac{i\alpha_j\xi(k^2-h^2)}{\alpha_j^2 k^2 - h^2\xi^2 + h^2k^2} \quad , \quad j = 1,2 \tag{8.11}$$

and

$$k^2 \equiv \rho\omega^2/\bar{\mu} \quad , \quad h^2 \equiv \rho\omega^2/(\bar{\lambda}+2\bar{\mu}) \tag{8.12}$$

The results (8.10) to (8.13) are identical to the classical forms with the exception that λ and μ are replaced by $\bar{\lambda}(\xi)$ and $\bar{\mu}(\xi)$. Substituting (8.10) into the boundary conditions, (8.4) gives a system of two homogeneous equations whose coefficient determinant must vanish. This gives the Rayleigh determinant

$$\left(\frac{k^2}{\xi^2} - 2\right)^4 = 16\left(1 - \frac{h^2}{\xi^2}\right)\left(1 - \frac{k^2}{\xi^2}\right) = 0 \tag{8.13}$$

whose root for ω/ξ is the Rayleigh wave speed. For a "nonlocal Poisson material," $(\bar{\lambda} = \bar{\mu})$, we have

$$c_R = \omega/\xi \approx 0.9194[\bar{\mu}(\xi)/\rho]^{1/2} \tag{8.14}$$

The case $\bar{\lambda} \neq \bar{\mu}$ can be obtained from (8.13) by numerical calculation.

While the form of (8.14) is the same as the classical one, we notice that $\bar{\mu}$ is a function of the wave number, and, therefore, these waves are *dispersive*. We now take for $\bar{\mu}$ the same functional form obtained in Section 7 for $\bar{\lambda}' + 2\bar{\mu}'$, i.e.,

$$\bar{\mu}'/\mu = \frac{4}{\xi^2 d^2} \sin^2 \frac{\xi d}{2} \tag{8.15}$$

which is valid only for the nearest neighborhood interactions. This gives

$$\frac{\omega}{\sqrt{\mu/\rho}} \simeq 1.8388 \left| \sin \frac{\xi d}{2} \right|, \quad \frac{-\pi}{d} \leq \xi \leq \frac{\pi}{d} \tag{8.16}$$

In Fig. 3 we give a plot of this and compare it with the results obtained from the atomic theory of lattices calculated by Gazis, et al. [1960] (see also Maradudin et al. [1971, p. 531]. As can be seen, the result is unbelievably good, thus indicating the great potential of the nonlocal theory in predicting the frequencies in the very short wave length region comparable to the atomic distances.

Note that this result is based on a model of nearest neighborhood interaction. For cases of far neighborhood interactions and for media having inhomogeneities, dislocations and local anisotropics, we still have available to us nonlocal theory, only in these cases the nonlocal kernels are not as simple as assumed here. Nevertheless, calculations are possible, at least in principle, for these cases for which the atomic lattice theory becomes too tedious, if not impossible, to handle.

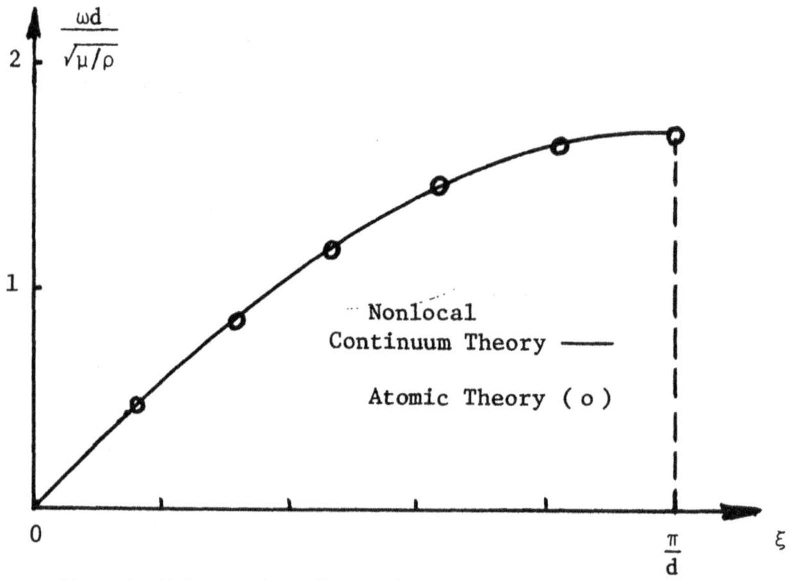

Fig. 3. Dispersion Curve for Surface Waves.

REFERENCES

Brillouin, L. [1946]: *Wave Propagation in Periodic Structures.* Dover.

Brockhouse, B. N. [1962]: *Phys. Rev.* 128, p. 1099.

Edelen, D. G. B. [1969]: *Arch. Rat. Mech. Anal.* 34, p. 283.

____________ and Laws, N. [1972]: *J. Math. Anal. Appl.* 38, p. 61.

Eringen, A. C. [1967]: *Developments in Mechanics* (ed. by T. Huang & M. Johnson). John Wiley & Sons, p. 23.

____________ [1967]: *Mechanics of Continua.* New York: John Wiley & Sons.

____________ [1972a]: "Nonlocal polar elastic continua," *Int. J. Engng. Sci.* 10, pp. 1-16.

____________ and Edelen, D. G. B. [1972b]: "On Nonlocal Elasticity, *Int. J. Engng. Sci.,* 10, pp. 233-248.

____________ [1972c]: "Linear Theory of Nonlocal Elasticity and Dispersion of Plane Waves," *Int. J. Engng. Sci.,* 10, pp. 425-435.

____________ [1972d]: "On Nonlocal Fluid Mechanics," *Int. J. Engng. Sci.* 10, pp. 561-575.

____________ [1973a]: "On Nonlocal Microfluid Mechanics," *Int. J. Engng. Sci.* 11, pp. 291-306.

____________ [1973b]: "Linear Theory of Nonlocal Microelasticity and Dispersion of Plane Waves," *Letters in Appl. & Engng. Sci.* 1, pp. 129-146.

____________ [1973c]: "Theory of Nonlocal Electromagnetic Elastic Solids," *J. Math. Phys.* 14, pp. 733-740.

____________ [1973d]: "On Rayleigh Surface Waves with Small Wave Lengths," *Letters in Appl. & Engng. Sci.* 1, p. 11.

____________ [1974a]: "Theory of Nonlocal Thermoelasticity," to appear in *Int. J. Engng. Sci.*

____________ [1974b]: "Memory Dependent Nonlocal Elastic Solids," to appear in *Letters in Appl. & Engng. Sci.*

____________ [1974c]: "On Nonlocal Continuum Thermodynamics," *Modern Developments in Thermodynamics* (ed. by B. Gal-Or). New York: John Wiley & Sons, pp. 121-142.

Gazis, D. C., Herman, R. and Wallis, R. F. [1960]: *Phys. Rev.* 119, p. 533.

Harrison, W. A. [1963]: *Phys. Rev.* 129, p. 2512.

Kröner, E. [1967]: *Int. J. Solids Struct.* 3, p. 731.

Kunin, I. A., edit. [1969]: *Mechanics of Generalized Continua.* Springer-Verlag, p. 321.

Maradudin, A. A., Montroll, E. W., Weiss, G. H. and Ipatava, I. P. [1971]: *Theory of Lattice Dynamics in the Harmonic Approximations,* 2nd ed., Academic Press.

Wallis, R. F., edit. [1963]: *Lattice Dynamics.* Pergamon Press.

ON THE PROBLEM OF CRACK TIP IN NONLOCAL ELASTICITY[1]

A. Cemal Eringen and B. S. Kim

Princeton University
Princeton, New Jersey 08540

ABSTRACT. In this article we present a brief account of our recent work[2] on the problem of stress distribution in a plate with crack under uniform tensile load at infinity perpendicular to the direction of crack. A fracture criterion is also given.

1. INTRODUCTION

In the previous article [2] presented in this volume, we developed the basic field equations of the nonlocal elasticity and applied them to the solution of the problems of propagation of plane and surface waves having wave lengths of any magnitude in the range between atomic distance and infinity. Here we present an outline of the solution of a classical problem in fracture mechanics namely: the stress concentration in the vicinity of a crack in an elastic plate subjected to a uniform tensile load at infinity perpendicular to the direction of crack. The classical elasticity solution of this problem is well known and it leads to infinite hoop stress at the crack tip. Since an infinite stress does not allow a yield or maximum stress hypothesis for fracture, other fracture criteria had to be introduced in the litterature circumventing the infinite stress. Starting with the celebrated fracture criterion of Griffith many different criteria were introduced which are based on the concepts of "stress intensity factor", stain energy, J-integral, etc.

1 This work was partially supported by the Army Research Office-Durham and the Office of Naval Research.
2 A paper under joint authorship with B. S. Kim will be published elsewhere; see also Eringen and Kim [1].

Thoft-Christensen (ed.), Continuum Mechanics Aspects of Geodynamics and Rock Fracture Mechanics, 107-113.

As we shall see the nonlocal elasticity solution of this problem leads to a finite stress at the crack tip, therefore, a fracture criterion based on the yield or maximum stress hypothesis can be restored. The results found are also interesting from the point of view of study of microcracks and other allied phenomena.

Indeed while it is always possible to "talk away" the crack tip problem by means of small curvature, plastic zones, cohesive forces, etc.,it appears to us that what is left is confusion and an insecure feeling. At any rate the solution given here can be looked upon as providing a solution of the crack problem for a perfect lattice (approximating brittle material). Perhaps more importantly, it opens a new and powerful avenue of thought for fracture mechanics requiring further intensive research.

2. THE CRACK PROBLEM

Here we outline briefly the solution of a simplified crack problem in a linear, homogeneous, isotropic, nonlocal elastic solid. Field equations to be solved are given by (6.6) of [2]. For the static case, vanishing body force ($\tilde{f} = \tilde{0}$) and no discontinuity surface this reads:

$$\int_V [(\lambda'+\mu')u'_{k,k\ell} + \mu' u'_{\ell,kk}]dv' = 0 . \tag{2.1}$$

We treat here only the plane strain problem. In this case the volume integral in (2.1) reduces to a surface integral. Further, we assume that the cohesive forces are of long range in the direction of crack only, and neglect the u_2-component of the displacement field in the crack direction since this effect is small in any case. Writing $v \equiv u_2$, $x_1 \equiv x$, $x_2 \equiv y$ then (2.1) reduces to

$$\int_{-\infty}^{\infty} [(\lambda'+2\mu') \frac{\partial^2 v(x',y)}{\partial y^2} + \mu' \frac{\partial^2 v(x',y)}{\partial x'^2}]dx' = 0 . \tag{2.2}$$

Here λ' and μ' are the nonlocal moduli which depend on $|x - x'|$. For a perfect one-dimensional lattice we have seen that (cf. 7.12 of [2])

$$(\lambda', \mu') = (\lambda, \mu)\alpha(|x - x'|) \tag{2.3}$$

where λ and μ are the Lamé constants and

$$\alpha(|x|) = \frac{1}{d}\left(1 - \frac{|x|}{d}\right), \text{ when } \frac{|x|}{d} < 1$$
$$= 0 \quad , \text{ when } \frac{x}{d} > 1 . \tag{2.4}$$

Clearly we need not use this expression of $\alpha(|x|)$ if some other function describes the nonlocal intermolecular forces better and this may be the case for imperfect solids.

For a pressurized crack with no other external forces the boundary conditions to be satisfied are:

$$v = 0 \quad \text{when } y = \infty$$
$$v = 0 \quad \text{when } y = 0, \ |x| \geq \ell \tag{2.5}$$
$$t_{yy} = -\sigma_o \quad \text{when } y = 0, \ |x| < \ell$$

where 2ℓ is the crack length. The solution of our problem, Fig. 2.1, is obtained by superposing to this solution the solution of the problem of plate, with no crack, subject to uniform tensile load $t_{yy} = \sigma_o$.

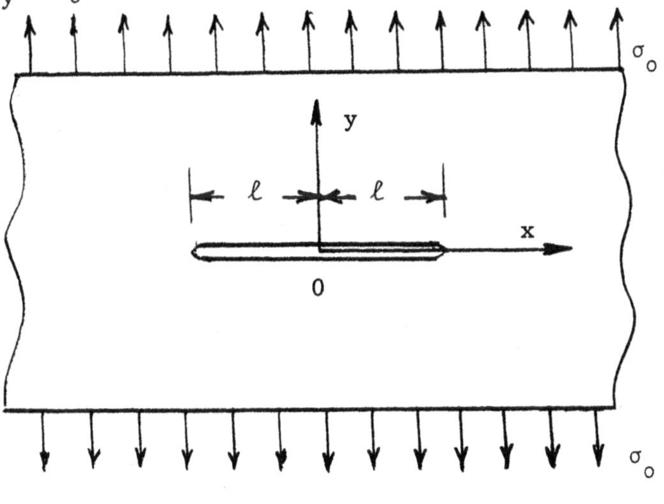

Fig. 2.1 Plate with Crack

The solution of (2.2) can be obtained by means of the Fourier transform

$$\bar{v}(\xi,y) \equiv \frac{1}{\sqrt{2\pi}} \int_{-\infty}^{\infty} v(x', y) \exp(i\xi x') \, dx' . \tag{2.6}$$

The Fourier transform of (2.2), using (2.3) gives

$$\frac{d^2\bar{v}}{dy^2} - \gamma^2\xi^2\bar{v} = 0 \text{ where } \gamma \equiv [\mu/(\lambda + 2\mu)]^{1/2} . \quad (2.7), (2.8)$$

The solution of (2.7) satisfying the boundary condition $(2.5)_1$, at the upper half plane at $y = \infty$, is

$$\bar{v}(\xi, y) = A(\xi) \exp(-\gamma|\xi|y), \; y \geq 0 . \quad (2.9)$$

The inverse Fourier transform gives

$$v(x, y) = \frac{1}{\sqrt{2\pi}}\int_{-\infty}^{\infty} A(\xi) \exp(-\gamma|\xi|y - i\xi x)d\xi . \quad (2.10)$$

The stress field t_{yy} and t_{yx} are now obtained through (6.5) of [2]:

$$\begin{aligned} t_{yy} &= -\gamma(\lambda+2\mu)\int_{-\infty}^{\infty} \bar{\alpha}(\xi) A(\xi)|\xi| \exp(-\gamma|\xi|y - i\xi x)d\xi \\ t_{yx} &= -i\mu \int_{-\infty}^{\infty} \bar{\alpha}(\xi) A(\xi)\xi \exp(-\gamma|\xi|y - i\xi x)d\xi . \end{aligned} \quad (2.11)$$

The remaining two boundary conditions $(2.5)_{2,3}$ are satisfied if

$$\begin{aligned} 0 &= \int_{-\infty}^{\infty} A(\xi) \exp(-i\xi x)d\xi, \; |x| \geq \ell \\ \sigma_0/\gamma(\lambda + 2\mu) &= \int_{-\infty}^{\infty} \bar{\alpha}(\xi) A(\xi)|\xi| \exp(-i\xi x)d\xi, \; |x| < \ell . \end{aligned} \quad (2.12)$$

For $\alpha(|x|)$ given by (2.4) we have obtained the solution of these dual integral equations. Leaving the details of our calculations to a separate publication I give some of the results here:

$$v(x, y) = \frac{\sigma_0 \ell}{\sqrt{\mu(\lambda+2\mu)}} \int_0^{\infty} \frac{J_1(\eta)}{\eta} \exp(-\frac{\gamma\eta y}{\ell}) \cos(\frac{x\eta}{\ell}) d\eta \quad (2.13)$$

$$\frac{t_{yy}(x, y)}{\sigma_0} = -(\frac{2\ell}{d})^2 \int_0^{\infty} \frac{J_1(\eta)}{\eta^2} \exp(-\frac{\gamma\eta y}{\ell}) \sin^2(\frac{d\eta}{2\ell}) \cos(\frac{x\eta}{\ell}) d\eta \quad (2.14)$$

where $J_1(\eta)$ is the Bessels function of order 1. Recall that for the plate under tension at infinity we must add 1 to the right-hand side of (2.14).

From (2.14) to a very high degree of accuracy (maximum possible error 0.2%) we obtain the stress concentration factor[1]

[1] For the case of tension load at infinity we must add 1 to the right-hand side of (2.15). However, when $2\ell >> d$ (2.15) can be used both in this case and in the case of pressurized crack.

$$\frac{\sigma}{\sigma_o} \simeq \frac{2}{3}\left(\frac{2\ell}{d}\right)^{\frac{1}{2}} \tag{2.15}$$

where we wrote $\sigma \equiv t_{yy}(\ell,0)$ for brevity.

3. FRACTURE CRITERION

(i) An examination of (2.13) to (2.15) shows that as $d \to 0$ the present results reduce to those obtained in the classical elasticity. Indeed in this case the nonlocal elasticity goes into classical elasticity. In fact the stress singularity according to (2.15) is a square root singularity which is well-known to us from the classical elasticity solution of the same crack problem.

(ii) Since the nonlocal elasticity leads to a finite stress at the crack tip we propose a fracture criterion based on the maximum stress.

Fracture Criterion. *When the tensile stress at the crack tip reaches a critical value σ_c the crack begins to expand.*

Note how simple, elegant and physically acceptable this result is. The question may be raised what this critical stress is. In the case of perfect lattices it is clear that σ_c is the theoretical yield stress in the crystal. It is well-known that this theoretical yield stress is 10^2 to 10^3 times the macroscopic yield stress [3]. In fact for steel if we use $\sigma_c \simeq E/5.5$ where E is Young's modulus and d = 3Å to calculate the fracture toughness K_{Ic} (as given by ASTM) we obtain

$$K_{Ic} = 41.9\ \sigma_o\sqrt{\ell} \simeq 26{,}300\ \text{psi}\sqrt{\text{in}} \tag{3.1}$$

where (2.15) is used with $\sigma = \sigma_c$. This result is in fair agreement with the experimental result $K_{Ic} \simeq 35{,}000$ psi$\sqrt{\text{in}}$ for steel at about -100°F [cf. ASTM, STP. 514, 1971, p. 166], considering the inaccuracy in d and σ_c based on macroscopic measurements and, of course, the factor 41.9 in (3.1) which is based on experiments on macrocracks.

If in (2.15) we take $\sigma = \sigma_c$ = the molecular yield stress, we obtain

$$\sigma^2\ell = \frac{9}{8}\sigma_c^2 d = \text{const.}\ . \tag{3.2}$$

This is exactly the same as the Griffith criterion for static fracture, except that here Griffith's constant is fully specified by $9\sigma_c^2 d/8$.

(iii) If the crack is subjected only to an internal pressure then the total applied load is $P_o = 2\sigma_o\ell$ = const.

If in this case $\sigma >> \sigma_c$, the crack will start propagating until it reaches the critical length $2\ell^*$ given by

$$2\ell^* = (2P_o/3\sigma_c)^2 d^{-1} \quad (3.3)$$

at which the applied pressure on the crack surface reduces to a value $P_0/2\ell^*$ and the propagation stops. Note that this is not the case for the tensile loading at infinity since in this case the total load is infinite.

(iv) For most steels we have an endurance limit in fatigue test; the ratio of tensile strength to fatigue limit is about $2 \sim 2.5$, c.f. Forrest [5]. Employing (2.15) this corresponds to $2\ell = 9d \sim 14d$, which implies that when the fatigue load is below the endurance limit, the crack length is few atomic distances. According to Francis [4] a microcrack of this length is not long enough to grow and become a macrocrack. Thus the nonlocal theory predicts also a minimum crack size below which no propagation can take place.

(v) Of course in real solids there exists a number of dislocations, vacancies and impurities so that the fatigue and growth of the crack can take place by accumulation of microcracks dislocations, etc. Clearly for real solids it is necessary to modify these specific results. But in nonlocal theory the kernel $\alpha(|x|)$ or more generally, $\sigma'_{k\ell mn}$ of (6.2) in [2] may be different. Nevertheless material internal-length parameters are introduced through the nonlocal elastic moduli which can be adjusted according to the nature of the solid.

In closing it is appropriate to make a precautionary remark in regard to all out grand expectation that the nonlocal theory will solve all of our outstanding problems in fracture mechanics. Clearly as every model the nonlocal theory too has limitations as is clear from the formulation given in [2]. The theory, however, contains the classical theory as a special case so that everything that can be done there is included here. But the converse is not true, and the nonlocal theory possesses a much larger domain of applicability, especially in problems where the long range interatomic forces become important. In this regard one need not even go to extremes such as atomic structure of materials. The theory is equally valuable for microscopic inner structures (e.g. granular solids, composites).

REFERENCES

[1] A. C. Eringen and B. S. Kim, "Stress Concentration at the Tip of Crack", Mechanics Communications, (1974).

[2] A. C. Eringen, "Nonlocal Elasticity and Waves", These proceedings.

[3] A. H. Cotrell, Dislocations and Plastic Flow in Crystals Oxford, 1953.

[4] P. H. Francis, "The Growth of Surface Microcracks in Fatigue," Trans. of ASME, J. Basic Engng. pp. 770-779, 1969.

[5] P. G. Forrest, Fatigue of Metals, Pergamon Press, New York, 1962.

STATISTICAL PROBLEMS IN THE THEORY OF ELASTICITY

E. Kröner

Institut für Theoretische und Angewandte Physik, Universität Stuttgart, Germany.

ABSTRACT. The present status of the theory of random elastic media is reviewed. A formal solution is derived which gives the tensor of the effective moduli in terms of the correlation functions up to infinite order of the distribution of the local elastic moduli. The formal solution is given in terms of multiple integrals which can be calculated only in favourable situations. Most important is the case of perfect disorder defined by a statistically independent distribution of the elastic moduli. In this case the integrals can be calculated and bounds which are correct to third order are derived. A peculiar difficulty which arises when the method is applied to dynamical problems is discussed.

1. INTRODUCTION

The subject of this lecture is the elasticity of bodies which are disordered on a microscopic scale. The most important examples of such bodies are polycrystalline aggregates and disordered composite materials, hence, in fact, those materials which mainly form the solid part of the earth. Therefore, a good theory of the effective, i.e. macroscopic elastic behaviour of such materials, should be valuable for gaining insight into

* This work has been supported by the Deutsche Forschungsgemeinschaft and the Stiftung Volkswagenwerk.

Thoft-Christensen (ed.), Continuum Mechanics Aspects of Geodynamics and Rock Fracture Mechanics, 115-134.

those phenomena in geophysics which have to do with the elastic properties of the earth.

Beside elastic also plastic and viscous deformation of the earth material is of great importance. Since both plastic and viscous deformation modes are always accompanied by elastic deformation, an elasticity theory of disordered materials will also be helpful in studying plasticity and viscosity. Whereas viscous deformation can often be treated in a linear theory, and therefore is accessible to similar methods as those used in elasticity theory, plastic deformation is basically a non-linear material property. As a consequence, the theory of plasticity is considerably more difficult than the theories of elasticity and viscosity. This statement applies to disordered media as well as to ordered media. It should be noted, however, that the statistical part of the theory of disordered plastic media is not basically different from that of the linear theories. Therefore, a study of the elasticity theory of disordered media can provide some guidance for the study of the plasticity of such media as well.

The elasticity theory of disordered media has made large progress in the last 10 years*. Certain basic problems have been solved finally. It is mainly these achievements about which I shall report in this lecture.

2. RECAPITULATION OF SOME NOTIONS OF PROBABILITY THEORY

Statistical physics is concerned with physical problems in which certain quantities of interest possess, at least partially, a random character. Due to this fact it is not possible to predict the outcome of an experiment in a deterministic way. The tossing of a die is a very simple example. However, if the experiment is repeated an increasing number of times one observes that the average of the outcomes approaches a value which can be predicted if certain informations are given about the problem. Such a value is called the expectation of the experiment. More generally, expectations are related not only to the average outcome of the experiment itself but also to more complicated results such as the average of the deviation-defined by some positive quantity - of the outcomes from the expection of this experiment. A fundamental law of

---

* See, for instance, refs.[1-2o]. This list does not claim completeness.

statistics states that the relative deviation which is the deviation normalized by the expectation tends to zero if the number of trials goes to infinity.

It is the declared aim of the statistical theories to calculate expectations from the statistical information about the physical situation. If the random character refers to spatial coordinates, as in statistical continuum mechanics where certain functions, called random or stochastic functions, fluctuate on a microscopic scale, the statistical information is also called macroscopic. An important theorem of statistics states that the calculation of expectations needs the complete statistical information. In the case of incomplete statistical information only bounds for expectations can be derived.

This theorem can easily be illustrated by another reference to the die-experiment. The full statistical information in this case is given by the probabilities for the outcomes 1,2,3,4,5,6. These probabilities are all equal to 1/6 if the faces of the die are equivalent in shape. The corresponding expectation is 3.5. It may be, however, that the die is less regularly built. For instance, one of it's corners may be protracted so that the probabilities of 1,2,3,4,5,6 are $p_1=p_2=p_3\neq p_4=p_5=p_6$. If the p's are not known we can, nevertheless, state that 2, the average of 1,2,3 should be a lower bound and 5, the average of 4,5,6 should be an upper bound to the expectation value. These bounds were the closest possible if the whole information were just $p_1=p_2=p_3 \neq p_4=p_5=p_6$. However, in our example the information means that the die has been built from a regular die by stretching out a corner along a space diagonal which means that the angles at the opposite corner remain 90°. Then we have the situation $p_1=p_2=p_3<p_4=p_5=p_6$. 2 and 5 are still bounds in this problem, but there are closer bounds, namely 3.5 and 5.

From these considerations it follows that the problem of statistics in the case of incomplete statistical information is to calculate the best-possible bounds for the expectations.

Of course, the averaging is to be weighted with the probabilities. These have the character of density functions if they are continuous-valued.

The set of experiments used for the averaging is also called an ensemble and the averages are ensemble averages. For the statistical continuum mechanics it is also common to think of ensembles as a group of systems similar to the physical system under consideration. Each member of the ensemble is prepared by the macroscopically equivalent procedure, hence only varies in the microscopic details. The expectation of a random function of position, such as the tensor $c(\underline{r})$ of elastic moduli, is then the average of its value at the corresponding points $\underline{r}$ in all systems of the ensemble. We denote this average by brackets <>. Thus $<c(\underline{r})>$ is the expectation of the tensor of elastic moduli at the point $\underline{r}$.

Expectations need not refer to a measurement at one point only. Of special importance are expectations of the form

$$\begin{aligned} <c_1c_2> &\equiv \int c_1c_2P_2(c_1c_2)dc_1dc_2 \\ <c_1c_2c_3> &\equiv \int c_1c_2c_3P_3(c_1c_2c_3)dc_1dc_2dc_3 \end{aligned} \tag{2.1}$$

etc., where $c_1 \equiv c(\underline{r}_1)$, P_2, P_3 etc. are the probability density functions of 2 points, 3 points etc. and the integrations extend over the range of the random function $c(\underline{r})$. The quantities defined by eqs.(2.1) are functions of $\underline{r}_1, \underline{r}_2$, of $\underline{r}_1, \underline{r}_2, \underline{r}_3$ etc. and are called n-point correlation functions if they refer to n points. If they depend on the relative positions of the points only the situation is called macroscopically (or statistically) homogeneous or stationary. If, however, they depend on the distances between the points only, the functions are macroscopically (or statistically) isotropic.

All functions, 1-point, 2-point etc., which arise from ensemble averaging are macroscopic functions i.e. they vary slowly on a spatial scale which is large compared to the scale of the random fluctuations. They can be differentiated and integrated. The hypothesis which makes these statements is the so-called ergodic hypothesis of the present theory. In order that this to be true there must exist the so-called representative, or macroscopic, volume elements ΔV which contain sufficient fluctuations of the random functions so that an average over ΔV is a meaningful operation. On the other hand, the ΔV must be small enough compared to the macroscopic dimensions in the problem so that differentiating and integrating on the basis of the ΔV becomes meaningful. If these conditions are satisfied in the considered physical situation then the ΔV-average $\int_{\Delta V} w(\underline{r})dV/\Delta V$ of a random 1-point function $w(r)$, as an example, is equal to the ensemble average $<w(\underline{r})>$ and the macroscopic integration over the whole medium gives

$$\int_V <w(\underline{r})>dV = \int_V w(\underline{r})dV \equiv W \tag{2.2}$$

where, for simplicity, we have replaced ΔV by dV on the left hand side. If $w(\underline{r})$ is the potential energy density then eq.(2.2) states that the total potential energy summed up over the average energies of the volume elements ΔV is equal to the true potential energy in the medium. This means that no energy is lost when passing from the microscopic deterministic description to the macroscopic statistical description of the medium.

Physical situations satisfy the above requirements if, and only if, they are "sufficiently random" in their character. We only consider situations in which this is the case.

In the theory to be outlined in the next section we make frequent use of the correlation functions. For simplicity we include the average $\langle c_1 \rangle = \langle c(\underline{r}_1) \rangle$ into the notion of correlation function. It can then be proved that the complete set of correlation functions of $c(\underline{r})$ which is characterized by n going from 1 to infinity contains the same information as does the complete set $P_1, P_2 \ldots\ldots P_\infty$ of the probability density functions of $c(\underline{r})$, i.e. the full statistical, or macroscopic, information about $c(\underline{r})$. Of course, the same applies to any random function of $\underline{r}$.

So far we have introduced the correlation functions in a rather formal way. They gain reality as soon as one thinks of devices of measuring them. There is no principal difficulty although the real measurement may be quite tedious. In a polycrystalline sample, for instance, a scanning electron microcope directed on a representative plane can detect the grain boundaries and orientations of the grains and from this information one can deduce the correlation functions.

The correlation functions exhibit the macroscopic features of the corresponding random function in a direct manner. For instance, the elastic moduli of aggregates which contain many grains elongated in some direction, are described by a 2-point correlation function which possesses a corresponding elongation in this direction. Boomerang-shaped grains show themselves in 3-point correlation functions etc.

3. THEORY OF THE EFFECTIVE ELASTIC MODULI OF DISORDERED MEDIA

The problem to be solved in this section consists in the calculation of the (4th rank) tensor C^{eff} of effective elastic moduli from the known statistical information about the distribution of the elastic moduli in the medium. This information will be given in the form of the correlation functions, introduced in section 2, of the tensor $c(\underline{r})$ of elastic moduli.

The tensor C^{eff} is defined by a macroscopic linear law of elasticity which relates the average stress with the average strain*:

$$\langle\sigma\rangle = C^{eff}\langle\varepsilon\rangle \ . \tag{3.1}$$

Depending on the distribution of the elastic moduli $c(\underline{r})$ this law is, in general, nonlocal. Therefore it has to be interpreted as

$$\langle\sigma(\underline{r})\rangle = \int_V C^{eff}(\underline{r},\underline{r}')\langle\varepsilon(\underline{r}')\rangle dV' \ . \tag{3.2}$$

The kernel $C^{eff}(\underline{r},\underline{r}')$ of eq.(3.2) has the range of the correlations of $c(\underline{r})$. A local law only exists when the range of the correlations vanishes in which case $C^{eff}(\underline{r},\underline{r}') = C(\underline{r}')\delta(\underline{r}-\underline{r}')$.

For simplicity, we shall restrict ourselves to macroscopically homogeneous distributions of $c(\underline{r})$. Then C^{eff} becomes a function of $\underline{r}-\underline{r}'$ in the nonlocal and a constant in the local situation.

In my first lecture at this symposium I had shown that

$$\varepsilon = \Gamma\sigma \tag{3.3}$$

is a solution of the equilibrium and compatibility equations of a medium with Hooke's law

$$\sigma = c\varepsilon \tag{3.4}$$

where Γ is the modified Green tensor of ref.[21] which can be adapted to particular boundary conditions. Since the Green tensor of an elastically inhomogeneous medium is not available it is desirable to reduce the problem to one of homogeneous elasticity with an (arbitrary) tensor $\mathring{C}$ of elastic moduli and a corresponding modified Green tensor $\mathring{\Gamma}$. This is achieved by inserting the previously introduced equation [21]

$$\Gamma = \mathring{\Gamma} - \mathring{\Gamma}\delta c\Gamma \ , \quad \delta c \equiv c-\mathring{C} \tag{3.5}$$

into eq.(3.3). Denoting $\mathring{\Gamma}\sigma$ by $\mathring{\varepsilon}$ we obtain the basic integral equation of the elasticity theory of inhomogeneous media in the form

* From now on we use the abridged notation described in ref.[21](this symposium).

$$\varepsilon = \mathring{\varepsilon} - \mathring{\Gamma}\delta c\varepsilon \quad . \tag{3.6}$$

According to P.H. Dederichs and R. Zeller [13] eq.(3.6) is formally equivalent to the celebrated Lippmann-Schwinger equation of the quantum mechanical scattering theory. Therefore, the methods for its solution developed there can be taken over to our present problem.

We note that eq.(3.5) applies to Green tensors $\Gamma, \mathring{\Gamma}$ which belong to the same type of boundary value problem, essentially the first and second problem and to that of the infinite medium.

Eq.(3.5) has the formal solution

$$\varepsilon = (I + \mathring{\Gamma}\delta c)^{-1}\, \mathring{\varepsilon} \tag{3.7}$$

in which the $(I+\mathring{\Gamma}\delta c)^{-1}$ is a multiple integral operator explained by the formal geometrical progression. Eqs. (3.6) or (3.7) are the physical equations of our problem which are now to be used in the statistical theory outlined in section 2.

According to eqs.(3.1) and (3.4) the required tensor C^{eff} connects the averages $\langle c\varepsilon\rangle$ and $\langle\varepsilon\rangle$. It can be shown [19] that $\mathring{\varepsilon}$ does not fluctuate. Therefore, the needed averages are

$$\langle\varepsilon\rangle = \langle(I+\mathring{\Gamma}\delta c)^{-1}\rangle\mathring{\varepsilon}, \quad \langle c\varepsilon\rangle = \langle c(I+\mathring{\Gamma}\delta c)^{-1}\rangle\mathring{\varepsilon} \quad . \tag{3.8}$$

If the first equation is solved for $\mathring{\varepsilon}$ and inserted in (3.7) we obtain after comparison with (3.1)

$$C^{eff} = \langle c(I+\mathring{\Gamma}\delta c)^{-1}\rangle\langle(I+\mathring{\Gamma}\delta c)^{-1}\rangle^{-1} \quad . \tag{3.9}$$

It is convenient, although not necessary, to set $\mathring{C}=\langle c\rangle$. Writing out the geometric progressions contained in (3.8) and putting $c-\langle c\rangle\equiv c'$ which implies $\langle c'\rangle=o$ we obtain a series of multiple integrals of increasing order:

$$\begin{aligned} C^{eff} = {} & \langle c\rangle - \langle c'\mathring{\Gamma}c'\rangle + \langle c'\mathring{\Gamma}c'\mathring{\Gamma}c'\rangle - \langle c'\mathring{\Gamma}c'\mathring{\Gamma}c'\mathring{\Gamma}c'\rangle \\ & + \langle c'\mathring{\Gamma}c'\rangle\mathring{\Gamma}\langle c'\mathring{\Gamma}c'\rangle + O(c'^5) \quad . \end{aligned} \tag{3.10}$$

Since the Green tensor $\mathring{\Gamma}$ is nonrandom it can be taken outside the averaging. Then the remaining averages are recognized as correlation functions of c', hence of c. The terms of order c'^n include correlation functions up to the order n.

From eq.(3.6) an equation of the form

$$\sigma = \mathring{\sigma} - \mathring{\Delta}\delta s\sigma \tag{3.11}$$

can easily be derived where $\overset{\circ}{\sigma} \equiv \overset{\circ}{\Delta}\varepsilon$ is nonrandom,

$$\overset{\circ}{\Gamma} = \overset{\circ}{S} - \overset{\circ}{S}\overset{\circ}{\Delta}\overset{\circ}{S}, \quad \delta s \equiv s - \overset{\circ}{S} \tag{3.12}$$

and s and $\overset{\circ}{S}$ are the tensors of elastic compliances of the real inhomogeneous and an arbitrary fictive homogeneous medium respectively. By exactly the same procedure as above the equation

$$\begin{aligned} S^{eff} = &\langle s\rangle - \langle s'\overset{\circ}{\Delta}s'\rangle + \langle s'\overset{\circ}{\Delta}s'\overset{\circ}{\Delta}s'\rangle - \langle s'\overset{\circ}{\Delta}s'\overset{\circ}{\Delta}s'\overset{\circ}{\Delta}s'\rangle \\ &+ \langle s'\overset{\circ}{\Delta}s'\rangle\overset{\circ}{\Delta}\langle s'\overset{\circ}{\Delta}s'\rangle + O(s'^5) \end{aligned} \tag{3.13}$$

is derived which we call the dual solution. Of course, the inverse of (3.13) gives the same result as (3.10) if all terms up to infinite order are included.

The dual solution becomes important when for computational reasons or for lack of information the series (3.10) is truncated. In this case the most valuable theorem of P.H. Dederichs and R. Zeller [14] applies according to which upper bounds to C^{eff} and S^{eff} are obtained if the series (3.10) and (3.13), respectively, are truncated after any terms of odd order. Upper bounds for S^{eff} transform into lower bounds for C^{eff} by inversion. It follows that $\langle c\rangle$ and $\langle s\rangle^{-1}$ are upper and lower bounds for C^{eff}, equal to the so-called Voigt and Reuss bounds, as pointed out first by R. Hill. We also call them first order bounds because they are correct to order O(c').

So-far no proposal has been made how to bound C^{eff} if the statistical information includes correlation functions up to an even order.

4. PERFECT DISORDER

Dederichs and Zeller [14] have also calculated 3rd-order bounds for the effective elastic (4th rank) tensor of polycrystalline aggregates of spherical grains. M. Hori & F. Yonezawa [18] use a cell model in order to calculate effective dielectric tensors and other material tensors of 2nd rank. In an earlier work I have introduced the notion of perfectly disordered materials. Here, perfect disorder is defined by a statistically independent distribution of the elastic moduli (or other material parameters). One can think of producing perfect disorder by tossing, separately for every point $\underline{r}$, a die which carries the possible values of the elastic moduli.

Perfect disorder implies, for instance, that the correlation function $\langle c_1 c_2\rangle$ vanishes if the points $\underline{r}_1$

and $\underline{r}_2$ are not the same. Certainly, this can only occur in an idealized sense because it requires a zero grain size. However, if the mean grain size is very small compared to the macroscopic dimensions of the specimen and if the orientation of the grains varies randomly from one grain to the other, then perfect disorder will reflect the physical situation very well.

In perfectly disordered media all n-point correlation functions are different from zero for a configuration of n-points if, and only if, there is no single point separated from all other points in the configuration. This follows from the statistical independence. In other words, the configurations which contribute to the integrals in the solution (3.10) consist of one or more conglomerates of at least two coinciding points. We shall describe the coincidence of 2 points, 3 points, 4 points etc. by a kind of generalized Kronecker symbols $\delta_{\underline{r}_{12}}$, $\delta_{\underline{r}_{123}}$, $\delta_{\underline{r}_{1234}}$ etc. which assume the value 1 if the points indicated coincide and are 0 otherwise. All δ's are totally symmetric in the subscripts. With this notation we can write down in a simple manner the correlation functions of c' needed in eq.(3.10). Up to 4th order we have

$$\langle c'_1 c'_2 \rangle = \underbrace{c'c'}\,\delta_{\underline{r}_{12}}$$

$$\langle c'_1 c'_2 c'_3 \rangle = \underbrace{c'c'c'}\,\delta_{\underline{r}_{123}} \tag{4.1}$$

$$\langle c'_1 c'_2 c'_3 c'_4 \rangle = \underbrace{c'c'c'c'}\,\delta_{\underline{r}_{1234}} + \underbrace{c'c'}\,\underbrace{c'c'}\,\delta_{\underline{r}_{12}}\delta_{\underline{r}_{34}}\gamma + \underbrace{c'c'c'c'}_{(13)(24)}\,\delta_{\underline{r}_{13}}\delta_{\underline{r}_{24}}\gamma + \underbrace{c'c'c'c'}_{(14)(23)}\,\delta_{\underline{r}_{14}}\delta_{\underline{r}_{23}}\gamma$$

where the symbols $\underbrace{\quad}$ $\underbrace{\quad\quad}$ $\underbrace{\quad\quad\quad}$ denote ensemble averaging and the γ's indicate that the two pairs of points must not coincide. We have used here these symbols because it is necessary to know which of the c' in the dyadic products are taken together in the averages. Another, more elaborate, way would be to write out the subscripts.

All averages on the right hand sides of eqs.(4.1) are autocorrelation functions, i.e. all c' in an average refer to one and the same point $\underline{r}$, say. Therefore, these averages are constant and can be set outside the integrals of eq.(3.10). Then the integrands contain only modified Green tensors and δ-symbols.

Because the correlation length is zero in the perfect disorder case it is permissible to use the

Green tensor of the infinite medium even if the medium is finite [13,21]. This tensor can be decomposed into two parts, namely

$$\overset{o}{\Gamma}=\overset{o}{\Gamma}_1+\overset{o}{\Gamma}_2 \equiv \overset{o}{E}\delta(\underline{r}-\underline{r}') + \overset{o}{F}/4\pi|\underline{r}-\underline{r}'|^3 \tag{4.2}$$

where the 4th rank tensors $\overset{o}{E}$ and $\overset{o}{F}$ are respectively constant and angle dependent only. $\overset{o}{F}$ changes the sign in such a way that its integration over the full space angle is zero. For this reason also the integrals

$\int dV_2 \ \overset{o}{\Gamma}_2(\underline{r}_1,\underline{r}_2)\delta_{\underline{r}_{12}}$, $\int dV_3 \ \overset{o}{\Gamma}_2(\underline{r}_2,\underline{r}_3)\delta_{\underline{r}_{123}}$,

$\int dV_4 \ \overset{o}{\Gamma}_2(r_3,r_4)\delta_{\underline{r}_{1234}}$ etc. vanish.

For isotropic $\overset{o}{\Gamma}$ we have

$$\overset{o}{E}_{ijkl}=\frac{1}{15\overset{o}{\mu}(3\overset{o}{\kappa}+4\overset{o}{\mu})}[-(3\overset{o}{\kappa}+4\overset{o}{\mu})\delta_{ij}\delta_{kl}+9(\overset{o}{\kappa}+2\overset{o}{\mu})I_{ijkl}] \tag{4.3}$$

$$\begin{aligned}\overset{o}{F}_{ijkl}=\frac{-1}{2\overset{o}{\mu}(3\overset{o}{\kappa}+4\overset{o}{\mu})}[&(3\overset{o}{\kappa}+\overset{o}{\mu})\delta_{ij}\delta_{kl}-6\overset{o}{\mu}I_{ijkl}-3(3\overset{o}{\kappa}+\overset{o}{\mu})\times\\ &\times(e_ie_j\delta_{kl}+e_ke_l\delta_{ij})-\tfrac{3}{2}(3\overset{o}{\kappa}-2\overset{o}{\mu})\times\\ &\times(e_je_l\delta_{ik}+e_ie_l\delta_{jk}+e_je_k\delta_{il}+e_ie_k\delta_{jl})\\ &+15(3\overset{o}{\kappa}+\overset{o}{\mu})e_ie_je_ke_l].\end{aligned} \tag{4.4}$$

Here, $e_i \equiv (x_i-x_i')/|\underline{r}-\underline{r}'|$ and $\overset{o}{\kappa}, \overset{o}{\mu}_o$ are the compression and shear modulus of the tensor $\overset{o}{C}$. We now insert the decomposition (4.2) in eq. (3.10). The 2nd order contribution is $-\langle c'\overset{o}{\Gamma}_1c'\rangle-\langle c'\overset{o}{\Gamma}_2c'\rangle$. The second term contains $\int dV_2\overset{o}{\Gamma}_2(\underline{r}_1,\underline{r}_2)\delta_{12}$ and therefore vanishes. The integration involved in the first term leads to $-\langle c'(\underline{r})\overset{o}{E}c'(\underline{r})\rangle$ where we have set $\underline{r}_1=\underline{r}$ for the simplicity of notation.

The 3rd order contribution is $\langle c'\overset{o}{\Gamma}_1c'\overset{o}{\Gamma}_1c'\rangle + \langle c'\overset{o}{\Gamma}_2c'\overset{o}{\Gamma}_1c'\rangle+\langle c'\overset{o}{\Gamma}_1c'\overset{o}{\Gamma}_2c'\rangle+\langle c'\overset{o}{\Gamma}_2c'\overset{o}{\Gamma}_2c'\rangle$. The first term results in $\langle c'(\underline{r})\overset{o}{E}c'(\underline{r})\overset{o}{E}c'(r)\rangle$. In the second and third term we put $c'\overset{o}{\Gamma}_1c'=\langle c'\overset{o}{\Gamma}_1c'\rangle+(c'\overset{o}{\Gamma}_1c')'$ and find that they vanish.* The last term contains $\int dV_3\overset{o}{\Gamma}_2(r_2,r_3)\delta_{123}$ and therefore vanishes.

The 4th order contribution is $\langle c'\overset{o}{\Gamma}_1c'\overset{o}{\Gamma}_1c'\overset{o}{\Gamma}_1c'\rangle$

$+\langle c'\overset{o}{\Gamma}_2 c'\overset{o}{\Gamma}_1 c'\overset{o}{\Gamma}_1c'\rangle+\langle c'\overset{o}{\Gamma}_1c'\overset{o}{\Gamma}_2c'\overset{o}{\Gamma}_1c'\rangle+\langle c'\overset{o}{\Gamma}_1c'\overset{o}{\Gamma}_1c'\overset{o}{\Gamma}_2c'\rangle$

* Throughout the paper, the prime denotes the deviation from the mean.

$+\langle c'\mathring{\Gamma}_2 c'\mathring{\Gamma}_2 c'\mathring{\Gamma}_1 c'\rangle + \langle c'\mathring{\Gamma}_2 c'\mathring{\Gamma}_1 c'\mathring{\Gamma}_2 c'\rangle + \langle c'\mathring{\Gamma}_1 c'\mathring{\Gamma}_2 c'\mathring{\Gamma}_2 c'\rangle$ $+\langle c'\mathring{\Gamma}_2 c'\mathring{\Gamma}_2 c'\mathring{\Gamma}_2 c'\rangle$. The first term is $\langle c'(\underline{r})\mathring{E}c'(\underline{r})\mathring{E}c'(r)\mathring{E}c'(r)\rangle$. Setting $c'\mathring{\Gamma}_1 c' = \langle c'\mathring{\Gamma}_1 c'\rangle + (c'\mathring{\Gamma}_1 c')'$ and $c'\mathring{\Gamma}_1 c'\mathring{\Gamma}_1 c' = \langle c'\mathring{\Gamma}_1 c'\mathring{\Gamma}_1 c'\rangle + (c'{}^{\circ}{}_1 c'{}^{\circ}{}_1 c')'$ we see that the 2nd, 3rd, 4th, 5th and 7th term vanish. The 6th term becomes equal to $\langle c'\mathring{\Gamma}_2 \langle c'\mathring{\Gamma}_1 c'\rangle \mathring{\Gamma}_2 c'\rangle$ whereas the 8th term has to be calculated as it stands.

With the obtained results sofar we can write C^{eff} up to 4th order as

$$\begin{aligned} C^{eff} &= \langle c\rangle - \langle c'(\underline{r})\mathring{E}c'(\underline{r})\rangle + \langle c'(\underline{r})\mathring{E}c'(\underline{r})\mathring{E}c'(r)\rangle \\ &\quad -\langle c'(\underline{r})\mathring{E}c'(\underline{r})\mathring{E}c'(\underline{r})\mathring{E}c'(\underline{r})\rangle \\ &\quad +\langle c'(\underline{r})\mathring{E}c'(\underline{r})\rangle\mathring{E}\langle c'(\underline{r})\mathring{E}c'(\underline{r})\rangle \qquad (4.5) \\ &\quad -\langle c'(\underline{r})\mathring{\Gamma}_2\langle c'(\underline{r})Ec'(\underline{r})\rangle\mathring{\Gamma}_2 c'(\underline{r})\rangle \\ &\quad -\langle c'(\underline{r})\mathring{\Gamma}_2 c'(\underline{r})\mathring{\Gamma}_2 c'(\underline{r})\mathring{\Gamma}_2 c'(\underline{r})\rangle \ . \end{aligned}$$

We remark that the terms which do not contain $\mathring{\Gamma}_2$ are the first terms of the geometric progression formed from the expression $\langle c(I+\mathring{E}c')^{-1}\rangle\langle(I+\mathring{E}c')^{-1}\rangle$ which follows from the r.h.s. of eq.(3.9) by the substitutions $\mathring{\Gamma} \to \mathring{E}, \delta c \to c'$.

The integrals containing $\mathring{\Gamma}_2$ are more involved although only elementary integrations occur. The auxiliary formula

$$\int dV_3\ \mathring{\Gamma}_2(\underline{r}_2,\underline{r}_3)\delta_{13} = \lim_{\rho\to o} \frac{4\pi\rho^3}{3}\ \mathring{\Gamma}_2(\underline{r}_2,\underline{r}_1) \qquad (4.6)$$

is of great help. It applies provided the point $\underline{r}_2$ lies outside a sphere of radius ρ around point $\underline{r}_1$, as is rather obvious.

We have not yet performed the integrations for the elastic medium where c' and $\mathring{\Gamma}$ are tensors of 4th rank. Although these calculations are just routine they turn out to be rather tedious. If one considers material properties which are described by a tensor of 2nd rank, such as dielectric and magnetic response or conductivity, then, after reinterpretation of c' and $\mathring{\Gamma}$, we are concerned with tensors of 2nd rank only. For this case we have performed all integrations up to 4th order. The results shall be given elsewhere.

I have treated the contributions of 4th order in some detail because among all non-vanishing terms which contain $\mathring{\Gamma}_2$ and, therefore, possess their specific problems the 4th order terms are of the lowest order. It is now clear that in the situation of perfect disorder all integrations involving $\mathring{\Gamma}_2$ can be performed, at least, in principle. This confirms the consistency of the concept of perfect disorder and its description by means of the δ-symbols. The most important feature of this concept is that it does not refer to particular grain shapes. By the use of the δ-symbols, it only requires that the mean grain shape be a sphere. Therefore, the concept of perfect disorder can also be applied to disordered media which are macroscopically anisotropic such as polycrystals with texture if they have no distinguishable grain shapes.

In this case one needs, beside the statement of perfect disorder, the (non-statistical) macroscopic information about the preferred orientations which is often given in terms of generalized spherical harmonics (see H.J.Bunge [22]). This is just the information needed to form averages like $\langle c\rangle, \langle c'_1 c'_1\rangle, \langle c'_1 c'_1 c'_1\rangle$ etc. If $\langle c\rangle$ which is now anisotropic, is calculated the tensor $\mathring{\Gamma}$ can be found. In the self-consistent theory of elastic moduli (see section 5) developed independently by A.V.Hershey [23] and E.Kröner [24] (see below) the case of texture was already considered and the part $\mathring{E}$ of $\mathring{\Gamma}$ has been calculated in important special cases by G.Kneer [25], P.R.Morris [26] and J.W.Hutchinson [27].

5. BOUNDS FOR EFFECTIVE ELASTIC MODULI AND SELF-CONSISTENT SCHEME

Everything we have said about the solution of eq.(3.6) is valid correspondingly for eq.(3.11), too. Here we write

$$\mathring{\Delta} = \mathring{\Delta}_1 + \mathring{\Delta}_2 \equiv \mathring{U}\delta(\underline{r}-\underline{r}') + \mathring{V}/4\pi|\underline{r}-\underline{r}'|^3 \qquad (5.1)$$

where $\mathring{U}$ is constant and $\mathring{V}$ is angle dependent only. For isotropic $\mathring{\Delta}$, formed with $\mathring{S}=\mathring{C}^{-1}$, we have

$$\mathring{U}_{ijkl} = \mathring{C}_{ijkl} - \mathring{C}_{ijmn}\mathring{E}_{mnpq}\mathring{C}_{pqkl} \qquad (5.2)$$

$$\mathring{V}_{ijkl} = -\mathring{C}_{ijmn}\mathring{F}_{mnpq}\mathring{C}_{pqkl} \qquad (5.3).$$

where $\mathring{E}$ and $\mathring{F}$ are to be taken from eqs.(4.3) and (4.4). For perfect disorder the series (3.10) and (3.13) become, up to 3rd order,

$$C^{(3')} = \langle c\rangle - \langle c'\overset{\circ}{E}c'\rangle + \langle c'\overset{\circ}{E}c'\overset{\circ}{E}c'\rangle \qquad (5.4)$$

$$S^{(3')} = \langle s\rangle - \langle s'\overset{\circ}{U}s'\rangle + \langle s'\overset{\circ}{U}s'\overset{\circ}{U}s'\rangle \,. \qquad (5.5)$$

As mentioned before these are upper bounds to C^{eff} and S^{eff} respectively. According to Dederichs and Zeller [14] one obtains (slightly) better bounds if all terms which do not contain $\overset{\circ}{\Gamma}_2$ or $\overset{\circ}{\Delta}_2$ are summed up to infinite order. We then obtain, valid for cubic c only,

$$C^{(3)} = \langle c(I+\overset{\circ}{E}c')^{-1}\rangle \, \langle (I+\overset{\circ}{E}c')^{-1}\rangle^{-1} \qquad (5.6)$$

$$S^{(3)} = \langle s(I+\overset{\circ}{U}s')^{-1}\rangle \, \langle (I+\overset{\circ}{U}s')^{-1}\rangle^{-1} \,. \qquad (5.7)$$

Also these bounds are of 3rd order because they deviate from the true C^{eff} and S^{eff} by terms starting at 4th order, namely the terms containing $\overset{\circ}{\Gamma}_2$ and $\overset{\circ}{\Delta}_2$. Eqs.(5.6) and (5.7) represent the best bounds which sofar have been evaluated for perfectly disordered bodies. It has not yet been proved, however, that they are the best possible bounds of 3rd order. For the case of statistically isotropic polycrystals with grains of cubic symmetry, only the effective shear modulus G^{eff} needs to be calculated because the compression modulus K^{eff} equals the orientation-independent compression modulus of the grains. The upper and lower bounds, $G^{(3)}$ and $G_{(3)}$, are

$$2G^{(3)} = 2\bar{\mu}\left(1-\frac{6\alpha\phi^2}{1-\alpha\phi}\right), \quad (2G_{(3)})^{-1} = 2\bar{m}\left(1-\frac{6\beta\psi^2}{1-\beta\psi}\right) \qquad (5.8)$$

where

$$\alpha \equiv \frac{2}{5}\cdot\frac{3\bar{\kappa}+6\bar{\mu}}{3\bar{\kappa}+4\bar{\mu}}\,, \quad \beta \equiv \frac{2}{5}\cdot\frac{3\bar{k}+6\bar{m}}{3\bar{k}+4\bar{m}}$$

$$\phi \equiv (\mu-\nu)/5\bar{\mu}, \quad 5\bar{\mu}=3\mu+2\nu, \quad \psi\equiv(m-n)/5\bar{m}, \quad 5\bar{m}=3m+2n$$

$$\kappa \equiv (c_{11}+2c_{12})/3, \quad \mu\equiv c_{44}, \quad \nu\equiv(c_{11}-c_{12})/2$$

$$k \equiv 4(s_{11}+2s_{12})/3, \quad m\equiv s_{44}/4, \quad n\equiv(s_{11}-s_{12})/2 \,.$$

c_{ij} and s_{ij} are respectively the Voigt components of the tensors c and s. The equations $\phi=0$ and $\psi=0$ imply isotropy, therefore ϕ and ψ are measures of the anisotropy of the grains in the polycrystal.

It is a remarkable fact that the so-called self-consistent elastic tensor C^{sc} of a perfectly disordered material always lies between the mentioned bounds of 3rd order. C^{sc} is calculated by solving a boundary value problem for a spherical inclusion in a homogeneous elastic matrix where c and C^{sc} are the elastic tensors of inclusion and matrix respectively. If the applied strain (and stress) is homogeneous in the region far away from the inclusion then the strain (and stress) in the inclusion is constant and dependent on c and C^{sc}. The condition of self-consistency is now that the strain (and stress) in the inclusion when averaged over all possible c(and s) equals the applied strain (and stress). In this way one obtains [24]

$$\langle \delta C(I+E^{sc}\delta C)^{-1} \rangle = 0 \tag{5.10}$$

$$\langle \delta S(I+U^{sc}\delta S)^{-1} \rangle = 0 \tag{5.11}$$

where $\delta C \equiv c-C^{sc}$, $\delta S = s-S^{sc}$, and E^{sc}, U_o^{sc} are similar to $\overset{o}{E}$ and $\overset{o}{U}$, however, C^{sc}, S^{sc} replacing $\overset{o}{C}$ and $\overset{o}{S}$. It can be seen that $C^{sc} = (S^{sc})^{-1}$. In fact, if this is true, eq.(5.11) is a consequence of eq.(5.10). This, however, can be shown easily by means of eq.(5.2) written with the self-consistent tensors. In particular, one has

$$(I + U^{sc}\delta S)^{-1} = c(I+E\delta C)^{-1}S \tag{5.12}$$

which effectively transforms eq.(5.10) into eq.(5.11).

We now argue that C^{sc} as determined from eq.(5.10) always lies below the upper bound $C^{(3)}$ of the perfectly disordered medium. Eq.(5.10) can be rewritten in a form which is very similar to eq.(5.6), namely

$$C^{sc} = \langle c(I+E^{sc}\delta C)^{-1} \rangle \langle (I+E^{sc}\delta C)^{-1} \rangle^{-1} . \tag{5.13}$$

Now $\delta C - c' = c - C^{sc} - (c - \langle c \rangle) = \langle c \rangle - C^{sc} > 0$ because $\langle c \rangle$ is an upper bound to C^{eff}. From the definition of $\overset{o}{E}$ one easily verifies that $E^{sc} > \overset{o}{E}$ the latter formed with $\langle c \rangle$. Hence $x \equiv E^{sc}\delta C > \overset{o}{E}c' \equiv y$. If eqs.(5.6) and (5.1) are written as power series in y and x respectively then the series have the same form and $C^{sc} < C^{(3)}$ follows from $x > y$. *

Similarly, one can show that $S^{sc} < S^{(3)}$ which means $C^{sc} > C_{(3)}$, the lower bound to C^{eff}. In this way one can prove our former statement that C^{sc} always lies between the bounds $C^{(3)}$ and $C_{(3)}$. This result makes C^{sc} particularly valuable as an approximation to the true effective tensor of elastic moduli of perfectly disordered materials. Clearly, C^{sc} is correct in the 3rd order of the fluctuations c', because $C^{(3)}$ and $C_{(3)}$ are so.

* Eqs. (5.6, 7) are proved for cubic c only. This restricts our result. As shown in a note dedicated to professor L. Sobrero (CISM, Udine 1974) C^{sc} is always bounded by $C^{(3')}$ and $C_{(3')}$.

6. WAVE PROPAGATION

For the sake of simplicity we confine ourselves to local effective elasticity in this section. A peculiar difficulty makes the treatment of wave propagation in disordered elastic systems more complicated than the statical problem dealt with sofar. In order to understand this let us first recall the definition of the effective elastic tensor given by eq.(3.1):

$$\langle\sigma\rangle = \langle c\varepsilon\rangle \equiv C^{eff}\langle\varepsilon\rangle \quad . \tag{6.1}$$

This definition leads to a macroscopic elastic energy density $\langle w\rangle = \langle\sigma\rangle\langle\varepsilon\rangle/2$. On the other hand the local elastic energy density is $\sigma\varepsilon/2$, hence $\langle w\rangle = \langle\sigma\varepsilon\rangle/2$. So the present theory gives

$$\langle\sigma\varepsilon\rangle = \langle\sigma\rangle\langle\varepsilon\rangle \quad , \tag{6.2}$$

a result which, at a first glance, looks rather strange. Eq.(6.2), with volume averages instead of ensemble averages, has been introduced by J.W.R. Bishop and R.Hill [28] in order to simplify their treatment of the plasticity of polycrystals. R. Hill [29] used it as the basis of his theory of the elastic behaviour of heterogeneous materials. Therefore, we shall call eq.(6.2) the Hill condition.

By means of the ergodic hypothesis it can be shown [15] that eq.(6.2) is, in fact, satisfied if the volume sources of the stresses, namely volume forces or incompatibilities*, are non-random. Since inertia forces are such volume forces we expect difficulties in connection with the Hill condition.

More directly, this can be seen also as follows: In the macroscopic theory, the kinetic energy density is expected to be proportional to $\langle v\rangle^2$ whereas it also must be proportional $\langle v^2\rangle$, if we take the mass density constant for simplicity. Of course, if the velocity v is random one always has $\langle v^2\rangle > \langle v\rangle^2$. Writing $v = \langle v\rangle + v'$ we obtain the result that the kinetic energy density contains a contribution proportional to $\langle v'^2\rangle$. Apparently, this implies that the mean wave field described by $\langle v\rangle$ is always accompanied by an extra motion derived from v' which carries a part of the kinetic and, as can be expected, of the potential energy. In a disordered material with constant mass density ρ,

---

* Volume incompatibilities are the sources of self-stresses, see ref.[21], this symposium.

for instance in a polycrystalline aggregate, we might try to determine the velocity of the mean wave from c^{eff} in the usual way. In fact, if we insert eq.(6.1) in the averaged wave equation

$$-\rho\frac{\partial^2}{\partial t^2}\langle u\rangle + \operatorname{div}\langle c\varepsilon\rangle = 0 \qquad (6.3)$$

where u is the displacement field producing the strain field ε we obtain

$$-\rho\frac{\partial^2}{\partial t^2}\langle u\rangle + \operatorname{div} c^{eff}\langle\varepsilon\rangle = 0 \,. \qquad (6.4)$$

If c^{eff} is isotropic than eq.(6.4) describes the propagation of the $\langle u\rangle$-field with sound velocities $\sqrt{(\lambda^{eff}+2\mu^{eff})/\rho}$ and $\sqrt{\mu^{eff}/\rho}$ where λ^{eff} and μ^{eff} are the Lamé constants of c^{eff}. In the above consideration we have used the fact that, due to the ergodic hypothesis, ensemble averaging and differentiating can be interchanged.

As mentioned above, the mean wave field must be accompanied by other motions. Equations for these can be established by multiplying the original wave equation by c', $c'c'$ etc. or by $c'\nabla$, $c'c'\nabla$ etc. and averaging the result. In this way one obtains equations of motion for correlation functions of the forms $\langle c'u\rangle$, $\langle c'c'u\rangle$ etc. or $\langle c'\varepsilon\rangle$, $\langle c'c'\varepsilon\rangle$ etc. . These equations are not pure wave equations but they couple among themselves and with the $\langle u\rangle$-field. The above correlation fields propagate through the medium thereby transporting the extra energy.

Eq.(6.4), if it is true, shows that the mean wave can propagate through the medium without damping losses. On the other hand, the multiple reflections which this wave suffers at the grain boundaries do, in fact, extract energy from the mean wave. The conclusion is that the mean wave obtains just as much energy from the correlation waves as it delivers to them via the reflections. In other words, one expects a dynamical equilibrium between the various types of waves.

These considerations are rather preliminary as yet. The conclusion about the dynamical equilibrium is in contradiction to a result of M.S. Howe [3o] who finds that "the net energy flux is always from the mean, or coherent, wave field to the random component of the wave field (in accordance with the second law of thermodynamics)". It may be questioned, however, whether the second law is applicable in our situation where the extra motion is not random but a propagation of correlation waves. Further investigation is necessary in order to clarify this point.

The situation is more involved if also the mass density is random, as in composite materials. In this case we might define an effective mass density by

$$\langle \rho \frac{\partial^2 u}{\partial t^2} \rangle \equiv \rho^{eff} \langle \frac{\partial^2 u}{\partial t^2} \rangle = \rho^{eff} \frac{\partial^2}{\partial t^2} \langle u \rangle \tag{6.5}$$

Instead of eq.(6.3) we obtain

$$\langle \rho \frac{\partial^2}{\partial t^2} u \rangle + \operatorname{div} \langle c\varepsilon \rangle = 0 \tag{6.6}$$

and this is equal to

$$-\rho^{eff} \frac{\partial^2}{\partial t^2} \langle u \rangle + \operatorname{div} C^{eff} \langle \varepsilon \rangle \ . \tag{6.7}$$

The two isotropic sound velocities are $\sqrt{(\lambda^{eff}+2\mu^{eff})/\rho^{eff}}$ and $\sqrt{\mu^{eff}/\rho^{eff}}$. To my knowledge, ρ^{eff} has not yet been calculated. In fact, it is not even clear that eq.(6.5) has a solution ρ^{eff} at all and that such a quantity, if it exists, is a scalar. A serious objection can be raised against the above treatment of wave propagation in random elastic media, especially against the derivation of eqs.(6.4) and (6.7). In order to understand this we reconsider the definition of C^{eff} once more. Eq.(6.1) implies that $\langle \varepsilon \rangle = 0$ leads to $\langle c\varepsilon \rangle = \langle \sigma \rangle = 0$. We now apply a well-known theorem of elasticity theory which states that by distributing suitable volume forces and incompatibilities as well as surface forces any strain field ε can be produced in any elastic medium, so also in our disordered material. Assume that in such a way we have produced a random field ε with $\langle \varepsilon \rangle = 0$. This means that at some points we have $\varepsilon > 0$ and at other points $\varepsilon < 0$. Now let c be distributed in such a way that $c < \langle c \rangle$ where $\varepsilon > 0$ and $c < \langle c \rangle$ where $\varepsilon < 0$. Then we have $\langle c\varepsilon \rangle = \langle \sigma \rangle > 0$ although $\langle \varepsilon \rangle = 0$. Clearly, eq.(6.1) is not applicable in such circumstances.

It is obvious that this situation contradicts the Hill condition and is, therefore, excluded when this condition is adopted. In fact, the l.h.s. of eq.(6.2) is always positive, if $\varepsilon \neq 0$, whereas the r.h.s. vanishes in our example. It is exactly the Hill condition which makes the definition (6.1) of the effective tensor C^{eff} permissible. As mentioned, the Hill condition is not valid if volume forces are present which are correlated to the local elastic moduli c. Therefore, the use of eq.(6.1) in the derivation of eq.(6.4) needs a justification which has not yet been given. Hence the problem of wave propagation in random elastic media is not yet solved.

7. SUMMARY

It was shown that the theory of random elastic media is in a good state as long as the volume forces and incompatibilities are non-random functions. In this case, a formal closed solution can be derived which expresses the effective elastic tensor C^{eff} of the medium in terms of the statistical information about the distribution of the local elastic moduli. The closed solution can be expended in a series of multiple integrals of increasing order which contain the statistical information in the form of correlation functions of the local moduli. If the series is truncated after any term of odd order one obtains an upper bound to C^{eff}. In a similar manner, lower bounds can be established.

Whether the required integrations can be performed depends on the distribution of the local moduli. Partucularly favourable and important is the statistically independent distribution, also known as perfect disorder, which makes it possible to perform all integrations. The bounds of 3rd order have been determined for aggregates of cubic crystals. They contain within themselves the self-consistent tensor C^{sc} which is, at the moment, the best approximation to the true C^{eff}.

The problem of wave propagation in random elastic media is less well understood. The reason is that here volume forces - the inertia forces - act which are correlated to the distribution of the elastic moduli. The complications which arise in this case are discussed and the direction in which further research is necessary is indicated.

ACKNOWLEDGEMENT

For numerous discussions I am grateful to my coworkers Dr. B.K. Datta-Gairola, Miss H. Koch, Mr. O.E. Eberhardt and Mr. K.-H. Krüger.

REFERENCES

1. I.M. Lifshitz and L.N. Rosenzweig, J.Exp.Teor. Fiz. 16, 967 (1946).
2. W.F. Brown, J.Chem.Phys. 23, 1514 (1955).
3. V.A. Lomakin, J.Appl.Math.Mech. 29, 1o48 (1965), translated from Russian.
4. S.D. Volkov and N.A. Klinskikh, Phys.Metals, Metallogr. 19, 24 (1965), translated from Russian.
5. M.J. Beran and J. Molyneux, Quart.Appl.Math. 24, 1o7 (1966).
6. E. Kröner, J.Mech.Phys.Solids 15, 319 (1967).
7. C. Eimer, Arch.Mech.Stos. 19, 4 (1967).
8. V.V. Bolotin and V.N.Moskalenko, Sov.Phys.Dokl. 13, 73 (1968).
9. M.J.Beran and J.J. McCoy, Int.J.Solids Struct. 6, 1o35 (197o)
10. V.V. Novoshilov, Prikl.Mat.Mech. 34, 67 (197o).
11. W.M. Lewin, Prikl.Mat.Mech. 35, 744 (1971).
12. P. Mazilu, Rev.Roum.Math.Pur.Appl. 17, 261 (1972).
13. R. Zeller and P.H. Dederichs, phys.stat.sol.(b) 55, 831 (1973).
14. P.H. Dederichs and R. Zeller, Z.Phys. 259, 1o3 (1973).
15. E. Kröner, Int.J.Eng.Sci. 11, 171 (1973).
16. J. Korringa, J.Math.Phys. 14, 5o9 (1973).
17. M. Hori, J.Math.Phys. 14, 514, 1942 (1973).
18. M. Hori and F. Yonezawa, J.Math.Phys. 15 (1974), in press.
19. E. Kröner, Statistical Continuum Mechanics, CISM Courses and Lectures No. 92, Udine 1971, Springer-Verlag Wien-New York, 1972.
20. A.G. Fokin and T.D. Shermergor, Transl. from PMM 32, 66o (1968).
21. E. Kröner, Internal stresses in crystals and in the earth, this symposium.
22. H.J. Bunge, Mathematische Methoden der Texturanalyse, Akademie-Verlag, Berlin, 1969.

23. A.V. Hershey, J.Appl.Mech. 21, 236 (1954).

24. E. Kröner, Z.Phys. 151, 5o4 (1958).

25. G. Kneer, phys.stat.sol. 9, 825 (1965).

26. P.R. Morris, Int.J.Eng.Sci. 8, 49 (197o).

27. J.W. Hutchinson, Proc.Roy.Soc. A 319, 247 (197o).

28. J.W.F. Bishop and R. Hill, Phil.Mag. 42, 414, 1298 (1951).

29. R. Hill, Proc.Phys.Soc. A 65, 349 (1952).

3o. M.S. Howe, Proc.Roy.Soc. A 331, 479 (1973).

INTERNAL STRESSES IN CRYSTALS AND IN THE EARTH

E. Kröner

Institut für Theoretische und Angewandte Physik, Universität Stuttgart, Germany

ABSTRACT. It is shown that the two important permanent deformation modes, namely plastic and viscous deformation, occur in similar form in crystals and in the earth. Earth quakes are compared with recovery phenomena in crystals and parallels are drawn between the sciences of the seismology in the earth and the acoustic emission in crystals. In a quantitative part the method of the modified Green functions is described which allows us to write complex results in a very condensed form. Hereby we confine ourselves to linear elasticity theory. The method also allows us to determine external and internal stresses in favourable situations.

1. INTRODUCTION

Among the physical properties of the solid part of the earth the mechanical properties are of particular importance. By mechanical we mean those properties which determine the deformation behaviour of the considered body. It is useful to classify these properties into elasticity, plasticity and viscosity. One of the latter two modes is always responsible for any permanent deformation, i.e. for a deformation which does not lead to the original configuration of the body if the forces producing the deformation are removed.

The division of the permanent deformation into plastic and viscous ones needs some convention. General agreement does not yet exist. The reason is probably

Thoft-Christensen (ed.), Continuum Mechanics Aspects of Geodynamics and Rock Fracture Mechanics, 135-150.

that, in general, the viscous flow in solids is accompanied by a plastic flow and vice versa. Often, however, the accompanying component is small so that it is convenient to speak of plastic or viscous deformation.

If a continuum standpoint is adopted one can employ the inherent or characteristic time criterion. A deformation which is independent of any such time is then classified as plastic. If one or more inherent times do exist, then the time-(or rate-)independent part can be separated and the remainder is classified as the viscous part of the deformation. R.S. Rivlin and A.C. Pipkin [1] have given this idea a beautiful mathematical formulation.

Solid state physicists try to explain why materials behave as they do and study the elementary processes on an atomic or microscopic scale. They find that no physical process is completely time-independent and, therefore, pure plasticity as defined above should not exist. They furthermore have discovered that the permanent deformation in solids occurs by the motion of line-shaped objects, the dislocations,in domains, where the particles (atoms, molecules) of the body are ordered (crystalline) and by the motion of point-shaped objects, the point defects, where the particles are in disorder. The findings are that the inherent times which determine the motion of the dislocations are so much smaller than the inherent times for the motion of the point defects that the former can often be set zero in a good approximation. With this convention it can be stated that the plastic flow takes place by the motion of (billions of) dislocations whereas the viscous flow is carried by (billions of) point defects.

A large part of the solid earth material is formed by polycrystalline minerals. These deform permanently both intracrystalline and intercrystalline. The first deformation mode is activated by dislocations whereas the second mode happens in the disordered boundary zones between adjacent grains by the disconnected jumping of single particles which act as point defects in this motion.* One speaks of grain boundary flow and this is a viscous deformation. The inherent time for this is pressure dependent and relatively large so that, except at elevated temperature, a relevant deformation

* More precisely one would say: the particle jumps to a free site leaving behind a free site. This means that the original free site, the point defect, has changed its place.

can be observed only over long or even very long periods. Nevertheless, this kind of deformation is very important since it can relax the great stresses in the earth.

The phenomena just described have been observed and investigated in the laboratories of the solid state physicists. Notwithstanding that the materials explored here were mainly metals one can assume that the main features which determine the character of these experiments emerge in our big laboratory, the earth, too, although the quantitative results will be somewhat different due to the difference in the materials involved and due to the large hydrostatic pressure which prevails in the earth.

It is the purpose of this lecture to report about results, methods and insights which have been gained in the study of crystal deformation and which may be useful in the study of the earth as well. This is the main content of section 2. The further sections deal with a quantitative method for determining external and internal stresses in elastic media, namely the method of the modified Green functions. Section 3 recapitulates some results concerning the ordinary Green tensor of elasticity theory. A basic formula which allows us to express the Green tensor of an inhomogeneous medium by the Green tensor of a homogeneous medium is derived. Using an abridged notation which is explained in section 4 we derive in section 5 a number of results about certain modified Green tensors which are of greatest importance for the treatment of many basic problems of elasticity theory. The method is applied to the problem of self-stresses in section 6.

2. DEFORMATION AND STRESSES IN CRYSTALS AND IN THE EARTH: A COMPARISON

The dislocation is defined as a topological peculiarity in an otherwise regular point lattice, the crystal. If the main interest lies in the stress field which surrounds the dislocation and in the deformation affected by the moving dislocation then it is good enough to define the dislocation as the line which bounds an internal limited surface S - often a plane - of the body along which a permanent relative displacement of the adjacent sides has occurred. If g is the vector, assumed constant over S, which describes this glide act then g is called the strength of the dislocation. The vector dL of the dislocation line element and the glide vector g determine the dislocation completely.

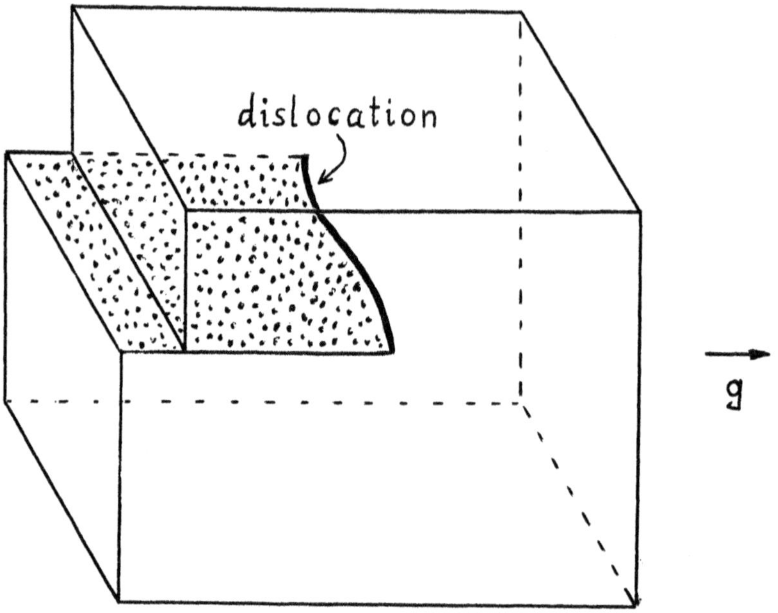

Fig. 1. Definition of the dislocation line.

In general, the dislocation, moving through a crystal, is stopped by some obstacle after it has moved a microscopic distance. Therefore, each moving dislocation implies a very localized two-dimensional glide.

Of course, the gliding is a deformation. External forces acting on the body can do work through this deformation. Thus we conclude that the dislocations have a tendency to move if the body is under load.

But how does the dislocation feel the presence of the external load? The natural answer is: through the stress field produced by this load. M.O. Peach and J.S. Koehler [2] have discovered that the (vector) line element dL of a dislocation with glide vector g in a (tensor) stress field σ experiences a force

$$dF = dL \times \sigma \cdot g \tag{2.1}$$

perpendicular to the line element. This result suggests that such a force also exists if the stress at the element dL is part of an internal stress field by which we mean a field of self-stresses (residual stresses). A dislocation would then tend to move in such a way that it can relax a part of the internal stress field. This is only possible if by the local slip of the dislocation a self-stress field is built up which has an opposite sign to the already existing stress field.

It is clear that such considerations are fundamental for the earth, too. The earth is in a state of stress which is caused by the gravitational and centrifugal forces, by an inhomogeneous temperature distribution and by preceeding plastic or viscous deformation. It is common in continuum mechanics to classify gravitational and centrifugal forces as external. The internal stresses would then be the temperature stresses and the residual stresses remaining after a removal of the gravitational and centrifugal forces, a procedure which, of course, cannot be achieved in reality.

Since the gravitational forces only act inside our "system" earth one can just as well include them into the cohesive forces of the earth. This would lead to a modified stress-strain law, namely a non-local law (see A.C. Eringen, this symposium) because the gravitational forces are long range. Whether this concept which accepts only the centrifugal forces as external or the former concept are better may depend on the intention. Both formulations of the theory are correct.

From the previous considerations follows that each local glide builds up a self-stress field which is opposite to the stress field which enforces the one in the glide zone concerned. In other words, each local glide leads to self-stresses which tend to stop further gliding in this zone. Elasticity theory states that these self-stresses extend to a considerable distance from the glide zone decaying as r^{-2} at a distance r from this zone.

The sign of the self-stresses outside the glide zone is such as to facilitate glide at other places. This fact can be elucidated by the example of the plastically deformed polycrystal. Assume a polycrystalline rod under simple tension. Since the plastic flow is restricted to certain crystallographic planes and directions not all grains will start gliding at the same time if the load is gradually increased. Rather there are a few grains of a most favourable lattice orientation in which gliding will set in whereas the other grains still behave as elastic. From this elastic environment the plastic grains experience an increasing back stress which hinders their further gliding.
At the same time self-stresses grow in the other grains

as a result of the preceeding plastic flow. These stresses are of the same sign as the applied stresses and, therefore, shift the start of plastic flow in these grains to lower applied stresses. Apparently, the tendency goes towards a plastic flow which is homogeneous throughout the whole polycrystal..

These considerations may bear some relevances to the problem of earth-quakes. There are regions in the earth where large shear stresses have been built up, usually during a long period of time. These stresses can partly relax by a glide act of the form described above. The glide zone will be limited due to the self-stresses produced by the gliding itself. It may be, however, that these self-stresses magnify the stresses already existing at other points to such an extent that also there glide can start. In this way the stresses can relax by several interconnected earth quakes.

A rather similar phenomenon is observed in crystals in which through their history, for instance by cold-working or heat treatment, large internal stresses have been built up. If such a piece of material is left alone in a suitable temperature range one often observes the so-called recovery which consists in a reduction of the internal stresses by the, more or less, correlated motion of dislocations or point defects. These motions are local glide acts as they also occur in earth quakes on a much larger scale.

As do earth quakes so also the local glide acts in crystals emit sound waves which propagate to the surface of the crystal. There they can be measured by suitable acoustic detectors. So in recent times a new branch of science has arisen called acoustic emission which deserves the attention of the geophysicists (see the contribution of S. Hanagud to this symposium). Acoustic emission is what also could be called crystal seismology. It is used in solid state physics in order to get information about the internal mechanical state of the body and about the processes which take place in it.

The examples sofar discussed may suffice to show that the deformation of crystals has much in common with the deformation of the earth's interior. It is possible to look at the crystal as a miniature model of earth in which on a small scale many of those phenomena occur which are observed on a large scale in real earth. In addition, most of the solid earth material is crystalline and therefore subject to the laws of crystal physics. This means, that it undergoes plastic and viscous deformation just as does the crystalline material in the laboratory. Hence the experimental motion of crystal dislocations and the grain boundary flow is of greatest importance also for the earth's interior.

Our considerations have been purely qualitative sofar. Great interest consists in quantitative predictions, for instance about future earth-quakes. Since the essential internal stresses of the earth vary slowly compared to the periods of time between earth quakes it seems to be possible to accumulate more and more information about the stress distribution in the earth interior. This information together with that of the material and temperature distribution seems to be indispensible for any reliable prediction.

As mentioned, the internal stresses are built up and relaxed by the plastic and viscous deformation in the earth interior. If sufficient information about this deformation is given one can calculate the internal stresses by means of continuum mechanics. This is only a minor part of the problem one really should solve, namely to calculate the whole flow of matter in the earth together with the accompanying stresses. Nevertheless, it is an important part which is solved in principle, at least. I shall, therefore, in the remainder of this lecture describe a general quantitative method for the determination of the internal stresses from the sources by which they are produced.

3. THE GREEN TENSOR OF ELASTICITY THEORY (RECAPITULATION)

In the following we restrict ourselves to linear elasticity theory. The method to be discussed is widely used in modern physics. It is called the method of the Green function, or rather the Green tensor, because in elasticity theory the Green function is tensor-valued with rank 2. It gives the displacement u_i at point $\underline{r}$ in the medium of volume V and surface S if the density of the volume forces degenerates to a point force of strength 1 at point $\underline{r}'$. Being a displacement the Green tensor has to satisfy the displacement equation of elasticity theory, namely

$$D_{ik}(\nabla)G_{km}(\underline{r},\underline{r}') + \delta(\underline{r}-\underline{r}')\delta_{im} = 0 \ . \qquad (3.1)$$

Here,

$$G_{km}(\underline{r},\underline{r}') = G_{mk}(\underline{r}',\underline{r}) \qquad (3.2)$$

are the cartesian components of the Green tensor and $D_{ik}(\nabla) \equiv \partial_j c_{ijkl}(\underline{r})\partial_l$ are the cartesian components of the differential operator of an elastic medium with elastic moduli c_{ijkl} which are allowed to depend on position. The appearance of the Dirac delta function $\delta(\underline{r}-\underline{r}')$ in eq. (3.1) takes care of the fact that the volume forces are degenerated to a point force whereas the Kronecker symbol δ_{im} expresses the fact that the point force has the strength 1.

Being an inhomogeneous differential equation for G_{km} eq. (3.1) does not define the Green tensor uniquely. Rather one may add any solution of the pertaining homogeneous equation to any particular solution of eq. (3.1). This is usually done in such a way that prescribed boundary conditions are satisfied. We shall only consider the first or displacement boundary value problem, the second or traction boundary value problem and the problem of the infinite medium. The respective Green tensors will be denoted by $G^{(1)}_{km}$, $G^{(2)}_{km}$ and $G^{(\infty)}_{km}$. The first two of them satisfy the boundary conditions

$$G^{(1)}_{km}(\underline{r},r')=0,\quad N_{ik}(\nabla)G^{(2)}_{km}(\underline{r},\underline{r}')+\delta_{im}/S=0,\quad \underline{r} \text{ on } S \tag{3.4}$$

where $N_{ik}(\nabla) = n_j c_{ijkl}(r)\partial_l$ and n_j is the outward unit vector normal to S.

Starting from the reciprocal theorem of E. Betti, C.Somigliana [3] and I. Fredholm [4] have derived the following identity for the elastic displacement field:

$$\begin{aligned} u_m(\underline{r}) = &- \int_V dV'G_{mk}(\underline{r},\underline{r}')D_{kl}(\nabla')u_l(\underline{r}') \\ &+ \int_S dS'G_{mk}(\underline{r},\underline{r}')N_{kl}(\nabla')u_l(\underline{r}') \\ &- \int_S dS'u_i(\underline{r}')N_{ik}(\nabla')G_{km}(\underline{r}',\underline{r}) . \end{aligned} \tag{3.5}$$

In the derivation of this formula in which u_i means the surface displacement it was assumed that the Green tensor satisfies eq. (3.1) but not necessarily any boundary condition. Hence (3.5) applies not only to $G^{(1)}$ and $G^{(2)}$ but also to $G^{(\infty)}$.

If we observe that the volume force density $F_k(\underline{r})$ and surface tractions $A_k(\underline{r})$ satisfy the equations

$$D_{kl}(\nabla)u_l(\underline{r}) = -F_k(\underline{r}),\quad N_{kl}(\nabla)u_l(\underline{r}) = A_k(\underline{r}) \tag{3.6}$$

we can use eq. (3.5) in order to express the displacement field by its sources, namely by F_k, A_k and u_i. We also note that due to the boundary conditions (3.4) the 2nd integral in eq. (3.5) vanishes if G_{mk} is taken as $G^{(1)}_{mk}$ and the 3rd integral vanishes with $G^{(2)}_{mk}$. If G^{∞}_{mk} is used in eq. (3.5) then, because the body is infinite, both the 2nd and the 3rd integral vanish, provided there are no stress sources at infinity.

It should be emphasized that eq. (3.5) and the just mentioned results are valid also when the body is elastically inhomogeneous. They can even be shown to apply to the non-local theory of elasticity if the operators $D_{kl}(\nabla')$ and $N_{kl}(\nabla')$ in eq. (3.5) are re-interpreted as $\partial_j \int dV'' c_{ijkl}(\underline{r}',\underline{r}'')\partial_l''$ and $n_j' \int dV'' c_{ijkl}(\underline{r}',\underline{r}'')\partial_l''$ respectively and $\underline{r}'$ is replaced by $\underline{r}''$ in the quantities on which these operators act. Because the Green tensor of such a body is usually not available one may, following G. Rieder [5], wish to treat the inhomogeneous problem by perturbation methods, in particular by the iterative solution of the corresponding problem in a homogeneous medium with an elasticity tensor $\overset{o}{c}_{ijkl}$, say. In particular, one may wish to express the Green tensor G_{km} of the inhomogeneous medium in terms of the simpler Green tensor $\overset{o}{G}_{km}$ of the homogeneous medium. The resulting formula is of great importance but little known in elasticity theory. It can be derived as follows: We set (omitting the subscripts)

$$c = \overset{o}{c} + \delta c, \quad D = \overset{o}{D} + \delta D, \quad N = \overset{o}{N} + \delta N, \quad G = \overset{o}{G} + \delta G \tag{3.7}$$

The eq. (3.1) for G_{km} can then be rewritten as

$$\begin{aligned} &\overset{o}{D}_{ik}(\nabla)\overset{o}{G}_{km}(\underline{r},\underline{r}') + \delta(\underline{r}-\underline{r}')\delta_{im} + \\ &+ \overset{o}{D}_{ik}(\nabla)\delta G_{km}(\underline{r},\underline{r}') + \delta D_{ik}(\nabla)G_{km}(\underline{r},\underline{r}') = 0 \end{aligned} \tag{3.8}$$

where the first two terms vanish, due to the definition of $\overset{o}{G}_{km}$. The corresponding boundary equation is

$$\overset{o}{N}_{ik}(\nabla)\delta G_{km}(\underline{r},\underline{r}') + \delta N_{ik}(\nabla)G_{km}(\underline{r},\underline{r}') = 0\ . \tag{3.9}$$

These equations are now solved for the 1st and 2nd boundary value problem and for the problem of the infinite body as well. To this end we consider the eqs. (3.8) and (3.9) as equations for the displacement $\delta G_{km}(\underline{r},\underline{r}')$ so that the Somigliana-Fredholm formalism can be applied (cf. eq. (3.5)). The result is in all three cases of the form

$$\delta G_{km}(\underline{r},\underline{r}') = \int_V dV'' \overset{o}{G}_{kn}(\underline{r},\underline{r}'')\partial''_p[\delta c_{nplq}(\underline{r}'')\partial''_q G_{lm}(\underline{r}'',\underline{r}')] \\ + \text{surface integrals.} \tag{3.10}$$

Partial integration gives

$$\delta G_{km}(\underline{r},\underline{r}') = -\int dV'' \overset{0}{G}_{kn,p''}(\underline{r},\underline{r}'')\delta c_{nplq}(r'')G_{lm,q''}(\underline{r}'',\underline{r}')$$
$$+ \text{ surface integrals } . \quad (3.11)$$

Here we have used the comma notation for differentiation.

In the case of the 1st problem the surface integrals are

$$\int dS'' \overset{0}{G}_{kn}(\underline{r},\underline{r}'')\ \delta N_{kl}(\nabla'')G_{lm}(\underline{r}'',\underline{r}')$$
$$-\int dS'' \delta G_{im}(\underline{r},\underline{r}'')\ \overset{0}{N}_{in}(\nabla'')\overset{0}{G}_{kn}(\underline{r}'',\underline{r}') \ .$$

They vanish because the Green tensors do so if $\underline{r}''$ is on S.

In the case of the 2nd problem the two surface integrals cancel each other. For the infinite body, finally, all surface integrals vanish because the displacement δG_{km} has no sources at infinity. So we have found that

$$\delta G_{km}(\underline{r},\underline{r}') = -\int_V dV'' \overset{0}{G}_{kn,p''}(\underline{r},\underline{r}'')\delta c_{nplq}(r'')G_{lm,q''}(r'',r') \quad (3.12)$$

in all three cases.

For later applications we now introduce a quantity of great importance which we call the modified Green tensor. It is of 4th rank and defined by

$$\Gamma_{ijkl}(\underline{r},\underline{r}') = G_{ik,jl'}(\underline{r},\underline{r}')\Big|_{(ij)(kl)} \quad (3.13)$$

where (ij) means symmetrization in j,j. Because of eq. (3.2) the tensor Γ is self-adjoint, i.e.

$$\Gamma_{ijkl}(\underline{r},\underline{r}') = \Gamma_{klij}(\underline{r}',\underline{r}) \ . \quad (3.14)$$

Differentiating eq. (3.12) with respect to x_i and x'_j and applying the symmetry operation (ik)(jm) we obtain

$$\delta\Gamma_{kimj} = -\int_V dV''\ \overset{0}{\Gamma}_{kinp}(\underline{r},r'')\delta c_{nplq}(r'')\Gamma_{lqmj}(r'',r'). \quad (3.15)$$

4. AN ABRIDGED NOTATION

Before we explain what we have achieved by introducing the modified Green tensor we describe, following R. Zeller and P.H. Dederichs [6], an abridged notation which makes the following considerations clearer. This notation uses the idea of the Hilbert space and is common in the quantum mechanical field theory. It makes it possible to write in a very compact form all equations which contain only symmetric 1-point tensors of the 2nd rank, like $\varepsilon_{ij}(\underline{r})$, and self-adjoint 2-point tensors of the 4th rank, like $\Gamma_{ijkl}(\underline{r},\underline{r}')$ where the tensor property refers to the physical, Euclidean, space E. Both kind of tensors will now be understood as quantities in the Hilbert space H, namely as vectors and as tensors of 2nd rank in H respectively. Volume integrations over two such quantities are then inner multiplications. We shall illustrate this by an example shortly. In order to be complete we still need the unit tensor I of the combined space (E,H). It is the product of the Euclidean unit tensor of 4th rank in E which has the components $(\delta_{ik}\delta_{jl}+\delta_{il}\delta_{jk})/2$ and the unit tensor of H which is $\delta(\underline{r}-\underline{r}')$.

In order to be consistent all tensors of 4th rank in E, so also the tensor c of elasticity, must be understood as 2-point functions. To this end we write the components of c, for instance, as $c_{ijkl}(\underline{r}')\delta(\underline{r}-\underline{r}')$. It is interesting to note that the whole formalism, including eq. (3.15) and the later main results, does also apply if the components of c are given in the more general form $c_{ijkl}(\underline{r},\underline{r}')$. This is the case of the non-local elasticity mentioned before. The law of elasticity, now written in the abridged notation as

$$\sigma = c\ \varepsilon\ , \tag{4.1}$$

then has the meaning

$$\sigma_{ij}(\underline{r}) = \int dV'\ c_{ijkl}(\underline{r},\underline{r}')\varepsilon_{kl}(\underline{r}')\ . \tag{4.2}$$

By comparison of eqs. (4.1) and (4.2) we also see the meaning of the product in the abridged notation. In the local case where the components of c become $c_{ijkl}(\underline{r}')\delta(\underline{r}-\underline{r}')$ eq. (4.2) degenerates to

$$\sigma_{ij}(\underline{r}) = c_{ijkl}(\underline{r})\varepsilon_{kl}(\underline{r})\ . \tag{4.3}$$

Also this equation is embodied in the short equation (4.1).

5. MAIN RESULTS OF THE THEORY OF MODIFIED GREEN TENSORS

We now return to eq. (3.15) and rewrite it in the short notation as

$$\Gamma = \overset{\circ}{\Gamma} - \overset{\circ}{\Gamma}\delta c\Gamma \ , \qquad \delta c = c - \overset{\circ}{c} \ . \tag{5.1}$$

This equation is formally identical with the celebrated Lippmann-Schwinger equation of the quantum mechanical scattering theory which means the methods developed there can be applied to problems of the inhomogeneous elasticity theory.

The importance of eq. (5.1) lies in the fact that with it we can express the complicated Green tensor Γ of the inhomogeneous problem in terms of the simpler $\overset{\circ}{\Gamma}$ of a corresponding homogeneous problem. In fact, eq. (5.1) can be solved for Γ to give

$$\begin{aligned} \Gamma &= (I + \overset{\circ}{\Gamma}\delta c)^{-1}\,\overset{\circ}{\Gamma} \\ &\equiv \overset{\circ}{\Gamma} - \overset{\circ}{\Gamma}\delta c\overset{\circ}{\Gamma} + \overset{\circ}{\Gamma}\delta c\overset{\circ}{\Gamma}\delta c\overset{\circ}{\Gamma} - \ldots\ldots + \ldots\ldots \end{aligned} \tag{5.2}$$

which is a development in terms of multiple integrals.

Before we explain which elastic problems are solved or can be solved if we know the modified Green tensor let us say a word on the compatibility equations. It was first noted by H. Reissner [7] that self-stresses are always associated with incompatible elastic strains. The compatibility equations of de St.Vernant can be written as

$$\text{Inc}\ \varepsilon = 0 \quad (-\varepsilon_{ikm}\varepsilon_{jln}\partial_k\partial_l\varepsilon_{mn}(\underline{r})=0) \tag{5.3}$$

(read: incompatibility of ε is zero). Developing further the ideas of Reissner we have to interpret the (symmetric) incompatibility tensor η, defined by

$$\text{Inc}\ \varepsilon = \eta \quad , \tag{5.4}$$

as a measure of the self-stresses. To solve a self-stress problem would then involve solving the incompatibility problem under the given boundary conditions.

It can be proved that a strain field in any elastic medium is uniquely determined if one prescribes either the set of quantities $(\eta, F, \overline{u})$ or the set (η, F, A), where F is the density of volume forces and $\overline{u}$, A are surface displacements and tractions respectively.

We now return to the modified Green tensor Γ. From the Somigliana-Fredholm formula and eqs. (3.6) one can easily derive that

$$\varepsilon = \Gamma^{(1)}\sigma \ , \quad (\eta = 0, \ \overline{u} = 0) \tag{5.5}$$

$$\varepsilon = \Gamma^{(2)}\sigma \ , \quad (\eta = 0) \tag{5.6}$$

where the superscripts (1) and (2) refer to the two boundary value problems. The eqs. (5.5) and (5.6) are valid under the conditions stated in parentheses. We interpret this result as follows. (i) If an elastic body which is free of self-stresses and the surface of which is fixed in space is deformed by volume forces then its stress and strain satisfy the eq. (5.5). (ii) If the body is free of self-stresses and deformed by volume forces and surface tractions then its stress and strain satisfy eq. (5.6).

These results show that $\Gamma^{(1)}$ and $\Gamma^{(2)}$ are projection operators. In fact, if any stress field σ is inserted on the r.h.s. of eqs. (5.5) and (5.6) then the strain field on the l.h.s. will only contain that part which can be attributed to a certain F or F,A respectively. In particular, the self-stress part perhaps contained in σ will be projected away. Thus, one can use $\Gamma^{(1)}$ and $\Gamma^{(2)}$ in order to produce finite or infinite series of strain fields which satisfy the indicated conditions and, therefore, can be used in variational procedures. This aspect of the Γ's has been discussed in ref. [8].

We now subtract the eqs. (5.5) and (5.6) from the equations expressing the inverse law of elasticity i.e. from $\varepsilon = s\sigma$, $s \equiv c^{-1}$ and obtain

$$\varepsilon = (s-\Gamma^{(1)})\sigma \qquad (F = 0) \tag{5.7}$$

$$\varepsilon = (s-\Gamma^{(2)})\sigma \qquad (F = 0, \ A = 0) \ . \tag{5.8}$$

After multiplication by c and with $\sigma = c\varepsilon$ these equations assume the form

$$\sigma = \Delta^{(1)}\varepsilon \ , \quad (F = 0) \tag{5.9}$$

$$\sigma = \Delta^{(2)}\varepsilon \ , \quad (F = 0, \ A = 0) \tag{5.10}$$

where the dual (modified) Green tensors Δ are given by

$$\Delta = c - c\,\Gamma\,c \ , \qquad \Gamma = s - s\,\Delta\,s \ . \qquad [5.11)$$

We interpret the above results as follows: (i) If an elastic body which is free of volume forces is deformed by incompatibilities and surface displacements then its stress and strain satisfy eq. (5.9).* (ii) If the body is free of volume forces and surface tractions but self-stressed then its stress and strain satisfy eq. (5.10).

Again, $\Delta^{(1)}$ and $\Delta^{(2)}$ are projection operators which annihilate strains which are due to forces F or F,A respectively.

The highly symmetric set of equations (5.5,6) and (5.9,10) are our main results of the theory of Green tensors in elastic media. These equations as well as eq. (4.1), the analogous equations

$$\Delta = \overset{\circ}{\Delta} - \overset{\circ}{\Delta}\delta s\Delta \quad , \quad \delta s \equiv s - \overset{\circ}{s} \qquad (5.12)$$

and the further equations

$$\Gamma = \Gamma\, c\, \Gamma \quad , \quad \Delta = \Delta\, s\, \Delta \qquad (5.13)$$

can be used even if the elasticity tensor c is anisotropic, inhomogeneous and/or non-local.

An important application of the formalism described in this section is to the theory of random elastic media treated in my second lecture at this symposium. In particular, eq. (5.1) will prove to be the basis in this theory.

In this section we have not mentioned the Green tensors $\Gamma^{(\infty)}$ and $\Delta^{(\infty)}$ of the infinite body. Of course, all formulas are valid for these tensors, too, if the situation excludes stress sources at infinity.

6. APPLICATION TO THE THEORY OF SELF-STRESSES

The dual (modified) Green tensor $\Delta^{(2)}$ annihilates strains which are produced by external volume and surface forces. These strains are compatible. Let us now assume that the internal stresses are due to some plastic or viscous strain for which we use the term extra strain ε^{ex}. The total strain ε^{tot} is then equal to $\varepsilon + \varepsilon^{ex}$, where ε is the elastic strain. Since ε^{tot} is compatible we find from eq. (5.10) that

$$\sigma = -\Delta^{(2)}\varepsilon^{ex} \quad , \quad (F = 0,\ A = 0)\ . \qquad (6.1)$$

This equation solves the problem of determining the

* It may sometimes be useful to think of surface incompatibilities. These would then be related to the surface displacements [9].

self-stresses of a source density ε^{ex} in the absence of external forces. It can also be applied to temperature and magnetostrictive stresses in which cases the extra strain requires a different interpretation.

Eq. (6.1) is equivalent to Maysel's formula (see W. Nowacki [10]) if this is written for the stress σ and for a stress-free surface, except that (6.1) applies also to inhomogeneous and non-local media.

Of course, the eq. (6.1) is useful only if the dual (modified) Green tensor $\Delta^{(2)}$ is known. According to its definition it can easily be obtained from the Green function $G^{(2)}$. If the medium is homogeneous, isotropic and infinite then the Green tensor $G^{(\infty)}$ has the Fourier components

$$\tilde{G}^{(\infty)}_{lm}(\underline{k}) = \frac{-(\lambda+\mu)k_l k_m + (\lambda+2\mu)k^2\delta_{lm}}{(2\pi)^{3/2}\mu(\lambda+2\mu)k^4} \tag{6.2}$$

where $\underline{k}$ is the wave vector. This formula can also be used if the above specified medium is non-locally elastic. The only modification is that the $\underline{k}$-dependent Lamé constants λ, μ have to be inserted.

If the situation is more complex the determination of the modified Green tensors becomes a difficult task. The problem then may require numerical methods such as the method of finite elements.

7. SUMMARY

Plastic and viscous deformation of solid matter leads to internal stresses. Through a number of examples it was shown that these phenomena are very similar in crystals and in the earth. In a way one may consider a crystalline body as a micro-earth which, for instance, suffers earth-quakes emitting seismic waves. Crystals resemble also in other aspects the big sister earth. Quantitative predictions are often desirable. They require quantitative methods, of course. Continuum mechanics is a valuable tool for this. A powerful method is that of the modified Green tensors which connect stress and strain in well-defined, general situations. Section 5 contains the most important equations for these tensors. Using them, one can solve general problems of external and internal stresses.* In principle, the theory is applicable also to elastic media which are anisotropic, inhomogeneous and non-local. Calculations then become much more complex and may require numerical methods except in very favourable situations.

---

* See J.D. Eshelby [11] for further quantitative connections.

REFERENCES

1. R.S. Rivlin and A.C. Pipkin, ZAMP, 16, 313, 1965.
2. M.O. Peach and J.S. Koehler, Phys.Rev., 80, 436, 1950.
3. C. Somigliana, Ann.Mat. (2), 17, 37, 1889.
4. I. Fredholm, Acta Math., 23, 1, 1900.
5. G. Rieder, Abh. Braunschw.Wiss.Ges., 14, 109, 1962.
6. R. Zeller and P.H. Dederichs, Phys.Stat.Sol. (b), 55, 831, 1973.
7. H. Reissner, ZAMM, 11, 1, 1931.
8. E. Kröner, Proc.Int.Symp. on Continuum Mechanics, Vienna, 1974, Springer-Verlag Wien-New York, to appear shortly.
9. E. Kröner, Erg.Ang.Math., 5, 1, 1958.
10. W. Nowacki, Thermoelasticity, Oxford, Pergamon Press, 1962.
11. J.D. Eshelby, Phil.Trans.Roy.Soc., A 274, 331, 1973.

THE ELEMENTS OF NON-LINEAR CONTINUUM MECHANICS

R.S. Rivlin

Lehigh University
Bethlehem, Pa., U.S.A.

1. INTRODUCTION

In this review, I shall give an introductory account of non-linear continuum mechanics, particularly emphasizing those aspects of the subject which may be of potential interest to geophysicists.

The most highly developed part of the subject is concerned with the mechanics of elastic materials which undergo deformations sufficiently large so that the classical theory of elasticity is not applicable. While many of the fundamental equations of this field were known in the nineteenth century, it was only about thirty years ago that serious progress was made in the solution of problems. At first the problems which were solved were restricted to the class of elastic materials which are isotropic and incompressible. With these restrictions, and no others, it was found possible to solve a number of problems of a rather simple character [1,2,3]: simple extension of a rod, biaxial deformation of a sheet, simple shear,simple torsion, and so on. Our ability to solve such problems resulted from the fact that in an isotropic incompressible material the deformation is in large measure kinematically determined. The importance of the solutions obtained in this way to the development of the subject of finite elasticity theory resides in the fact that they were obtained without further restrictions on the material than those implied by isotropy and incompressibility. For materials for which these restrictions are valid, they enabled us to characterize the materials by comparing experimental measurements with the results of the calculations. At the time, the classes of problems which could be solved were restricted by the non-linearity of the governing differential equations to those which possess a sufficiently high degree of symmetry to render them effectively one-dimensional, or to those

Thoft-Christensen (ed.), Continuum Mechanics Aspects of Geodynamics and Rock Fracture Mechanics, 151-175.

for which, for one reason or another, the governing differential equations could be linearized. More recently, the application of high-speed computers, using finite element methods, to the solution of problems in finite elasticity, particularly by Oden and by Klosner (see, for example [4]), has significantly increased the potentialities of the theory in the solution of specific problems.

Following the pioneering paper of Oldroyd [5] in 1950, the construction of a continuum-mechanical theory for non-linear visco-elastic materials, both solid and fluid, underwent a considerable development [6,7,8] in the decade to 1960. In the development of the basic equations of the theory, ideas already present and well-understood in the context of finite elasticity found an extension to a much more general class of materials. As might be expected, in view of the width of this class, the actual characterization of the materials, already very difficult in finite elasticity theory for compressible materials, acquires new dimensions of impracticality and has not to date been satisfactorily solved. Perhaps it never will be solved in a completely satisfactory manner. I hope that the second half of this paper will provide some insight as to why this may be the case.

While it is, of course, rash to predict how any theory will develop in the future, it may well be that the future of non-linear viscoelasticity theory as a tool in stress analysis will depend on the characterization of the material for restricted classes of deformations, such as deformations which are sufficiently slow, or sufficiently small, or have some specified form of dependence on time. Exceptions may possibly arise in the case of materials for which the constitutive equations can be seen from exploratory experiments to have a fortuitously simple form, or for which the constitutive equation can be derived from a structural model.

2. SOME KINEMATIC CONSIDERATIONS

Let X and x be the vector positions, with respect to a fixed origin, of a generic particle of a body at some fixed reference time T and at time t respectively. Then, if the relation

$$x = x(X,t) \tag{2.1}$$

is known, the deformation of the body is completely described. We shall regard the body as undeformed at the reference time T. The displacement vector u at time t is defined by

$$u = x - X\ . \tag{2.2}$$

Let x_i, X_A and u_i be the components in a rectangular cartesian coordinate system x of the vectors x, X and u respectively.

Then, the deformation gradient tensor g has components* $g_{iA} = x_{i,A}$ in the system x. Thus, we may write

$$g = \| g_{iA} \| = \| x_{i,A} \| . \tag{2.3}$$

The displacement gradient tensor has components $u_{i,A}$ in the system x.

For deformations which are possible in a real material,

$$\det g > 0 \tag{2.4}$$

and we note that $\det g$ is the ratio between the volume of a material element of the body in its deformed and undeformed states.

Since any real matrix may be written as the product of an orthogonal matrix r, say, and a symmetric matrix s, say, we have

$$g = rs , \tag{2.5}$$

whence it follows that

$$s^2 = g^T g = C , \quad \text{say,} \quad \text{and} \quad r = gC^{-\frac{1}{2}} , \tag{2.6}$$

where the superscript T denotes the transpose. C is called the Cauchy strain tensor and, in the cartesian coordinate system x, we may write its components C_{AB} as

$$C_{AB} = x_{i,A} x_{i,B} . \tag{2.7}$$

We may also decompose g into the product of an orthogonal matrix r and a symmetric matrix s in the inverse order. Thus,

$$g = sr . \tag{2.8}$$

Then,

$$s^2 = gg^T = c^2 , \quad \text{say,} \quad \text{and} \quad r = c^{-\frac{1}{2}} g . \tag{2.9}$$

c is called the Finger strain tensor and, in the coordinate system x, we may write its components c_{ij} as

$$c_{ij} = x_{i,A} x_{j,A} . \tag{2.10}$$

s and r defined by the decomposition (2.8) are, of course,

* We use the notation $,A$ to denote the operator $\partial/\partial X_A$ and the notation $,i$ to denote the operator $\partial/\partial x_i$.

generally different from those defined by the decomposition (2.5). However, it is easily shown that the principal values of C and c are the same and are all positive. It follows that the principal values of s given by (2.6) and (2.9) are the same and are real. They are, of course, determined by (2.6) and (2.9) apart from ambiguities of sign. If the sign of each of the principal values of s is taken to be positive, then it follows from (2.4) that r is, in either case, a *proper* orthogonal matrix. Also, since a symmetric matrix describes a pure homogeneous deformation, we can interpret (2.5) and (2.8) as the *polar decomposition theorem*: The deformation of an infinitesimal element of a body may be regarded as the resultant of a pure homogeneous deformation followed by a rotation or as the resultant of a rotation followed by a pure homogeneous deformation.

3. BASIC MECHANICAL EQUATIONS

We define the *Piola-Kirchhoff stress tensor* Π, with components Π_{Bi} in the coordinate system x, in the following way. Let $\Pi_1 dA$ be the force at time t, acting on an element of area in the body, which was, at time T, normal to the axis x_1 of the system x and had area dA at time T. Π_1 has components Π_{1i} in the system x. Similar definitions are given for Π_2 and Π_3 with components Π_{2i} and Π_{3i}. The components Π_{Ai} of the three vectors Π_1, Π_2 and Π_3 are those of the Piola-Kirchhoff stress tensor. They satisfy the *point equation of motion*

$$\Pi_{Ai,A} + \rho_0 \phi_i = \rho_0 \ddot{x}_i , \tag{3.1}$$

where ϕ_i are the components of the body force ϕ per unit mass, ρ_0 is the material density at time T, and a dot denotes material differentiation with respect to t. Also, if F_i are the components of the force F, at time t, acting on the surface of the body, per unit area measured at time T, then

$$F_i = \Pi_{Ai} N_A , \tag{3.2}$$

where N_A are the components in the system x of the unit normal N to the surface at time T. Π_{Ai} also satisfies the symmetry condition

$$x_{i,A}\Pi_{Aj} = x_{j,A}\Pi_{Ai} . \tag{3.3}$$

If we take the reference time T to coincide with time t , then the Piola-Kirchhoff stress becomes the Cauchy stress σ , with components σ_{ij} ; the surface force F becomes the surface force f , per unit area measured at time t , and equations (3.1), (3.2) and (3.3) become

$$\sigma_{ij,j} + \rho\phi_i = \rho\ddot{x}_i \ , \tag{3.4}$$

$$f_i = \sigma_{ji}n_j \ , \tag{3.5}$$

and

$$\sigma_{ij} = \sigma_{ji} \tag{3.6}$$

respectively, where ρ is the material density at time t and n_i are the components of the unit normal to the surface at time t .

It can be shown, from purely geometrical considerations, that the Piola-Kirchhoff and Cauchy stress tensors are related by the equation

$$\sigma = \frac{1}{\det g}\, g\, \Pi \ . \tag{3.7}$$

Also, the law of conservation of mass leads to the relation

$$\rho_0 = \rho \det g \ . \tag{3.8}$$

4. THE STRAIN-ENERGY FUNCTION FOR AN ELASTIC MATERIAL

For isothermal deformations of an elastic material, the specific Helmholtz free energy W (i.e. the Helmholtz free-energy per unit mass) and the Piola-Kirchhoff stress, at the instant of time t , are functions of the deformation gradient tensor at time t . Thus,

$$W = W(g) \ , \quad \Pi = \Pi(g) \ . \tag{4.1}$$

Then, it can be shown from the second law of thermodynamics that Π is given in terms of W by the relation

$$\Pi = \rho_0 \left(\frac{\partial W}{\partial g}\right)^T \ , \tag{4.2}$$

or, in cartesian notation, by

$$\Pi_{Ai} = \rho_0 \frac{\partial W}{\partial x_{i,A}} \ . \tag{4.3}$$

With (3.7), we obtain

$$\sigma_{ji} = \frac{\rho_0}{\det g} x_{j,A} \frac{\partial W}{\partial x_{i,A}} . \tag{4.4}$$

In deriving (4.3) and (4.4), it is assumed that there are no constraints on g, i.e. its components can be varied independently. This is not the case if the material considered is incompressible. Then, the relation (4.3) must be replaced by

$$\Pi_{Ai} = \rho_0 \frac{\partial W}{\partial x_{i,A}} - \frac{1}{2} p\varepsilon_{ijk}\varepsilon_{ABC}x_{j,B}x_{k,C} , \tag{4.5}$$

where ε_{ijk} is the alternating symbol and p is a hydrostatic pressure, which is undetermined if the deformation is given. Correspondingly, (4.4) must be replaced by

$$\sigma_{ji} = \rho_0 x_{j,A} \frac{\partial W}{\partial x_{i,A}} - p\delta_{ij} , \tag{4.6}$$

where δ_{ij} is the Kronecker delta.

W, which in mechanics is sometimes called the *strain-energy function*, cannot depend on g in an arbitrary manner. It is restricted by the consideration that it is unaltered if we superpose on the assumed deformation an arbitrary rigid rotation. Let $\bar{g}$ be the deformation gradient tensor for the resulting deformation so obtained. Then,

$$\bar{g} = rg , \tag{4.7}$$

where r is a proper orthogonal tensor. We have

$$W(g) = W(\bar{g}) . \tag{4.8}$$

This relation must be satisfied for all proper orthogonal r and in particular for $r = C^{-\frac{1}{2}}g^T$, thus

$$W(g) = W(C^{-\frac{1}{2}}g^T g) = W(C^{\frac{1}{2}}) . \tag{4.9}$$

We now note that C, defined by (2.6), is unchanged if g is replaced by rg, where r is an *arbitrary* proper orthogonal tensor. Accordingly, W must be expressible as a function of the Cauchy strain tensor C, thus

$$W = W(C) . \tag{4.10}$$

Introducing (4.10) into (4.3) and (4.4), we obtain

$$\Pi_{Ai} = \rho_0 x_{i,B}\left(\frac{\partial W}{\partial C_{AB}} + \frac{\partial W}{\partial C_{BA}}\right) \tag{4.11}$$

and

$$\sigma_{ji} = \frac{\rho_0}{\det g} x_{i,A} x_{j,B} \left(\frac{\partial W}{\partial C_{AB}} + \frac{\partial W}{\partial C_{BA}} \right) . \tag{4.12}$$

5. RESTRICTIONS DUE TO MATERIAL SYMMETRY

We shall now suppose that the elastic material possesses some symmetry in its reference configuration at time T. This is described by an appropriate group of transformations $\{S\}$ - the symmetry group of the material. Then, W must satisfy the relation

$$W(S\, C\, S^T) = W(C) , \tag{5.1}$$

for all transformations of the group $\{S\}$. The implicit restrictions imposed on W by (5.1) may be made explicit for any particular symmetry group by using the methods of the theory of invariants. Equation (5.1) expresses the fact that W is a scalar invariant of the symmetric tensor C under the transformation group $\{S\}$. If W is a polynomial in C, then it may be expressed as a polynomial in the elements of an irreducible integrity basis for the symmetric tensor C under the group $\{S\}$. If W is a function of C, then it may be expressed as a function of these elements.

For an isotropic material, the appropriate symmetry group is the full or proper three-dimensional orthogonal group and W can be expressed as a function of the invariants I_1, I_2, I_3, defined by

$$I_1 = \operatorname{tr} C , \quad I_2 = \tfrac{1}{2}\{(\operatorname{tr} C)^2 - \operatorname{tr} C^2\} , \quad I_3 = \det C , \tag{5.2}$$

which form an irreducible integrity basis for C under the full or proper three-dimensional orthogonal group.

For a material which is transversely isotropic (i.e. which possesses fiber symmetry), the appropriate symmetry group is the full or proper two-dimensional orthogonal group in the plane normal to the fiber axis. W can then be expressed [9] as a function of the invariants $I_1, I_2, \ldots, I_5$, where I_1, I_2 and I_3 are defined by (5.2) and I_4 and I_5 are defined by

$$I_4 = C_{33} , \quad I_5 = C_{3A}\, C_{A3} , \tag{5.3}$$

the fiber axis being taken as the x_3-direction of the reference system x.

For each of the thirty-two crystal classes, the corresponding results have been obtained by Smith and Rivlin [10]. They found that these thirty-two classes divide into eleven groups for each of which a different irreducible integrity basis is obtained.

Since, for an isotropic material W must be expressible in the form

$$W = W(I_1, I_2, I_3) , \tag{5.4}$$

where I_1, I_2 and I_3 are given by (5.2), we can obtain an expression for the Cauchy stress σ, by introducing (5.4) into (4.12) and using (5.2). We obtain

$$\sigma = \frac{2\rho_0}{I_3^{\frac{1}{2}}}[(W_1+I_1W_2)c - W_2c^2 + I_3W_3\delta] , \tag{5.5}$$

where δ is the identity tensor and the notation $W_\alpha = \partial W/\partial I_\alpha$ (α=1,2,3) is used. This is the constitutive equation for an isotropic elastic material.

In an incompressible material $\det g = 1$ for all possible deformations. It is easily shown that $\det C = (\det g)^2$. Accordingly,

$$I_3 = 1 . \tag{5.6}$$

It follows that for an incompressible isotropic elastic material, W must be expressible as a function of I_1 and I_2 , thus

$$W = W(I_1,I_2) . \tag{5.7}$$

Introducing (5.7) into (4.6) and using (5.2), we obtain [1]

$$\sigma = 2\rho_0[(W_1+I_1W_2)c - W_2c^2] - p\delta . \tag{5.8}$$

More generally, if for any specified symmetry W is expressible as a function of the invariants $K_1,K_2,\ldots,K_\lambda$ thus

$$W = W(K_1,\ldots,K_\lambda) , \tag{5.9}$$

we have, from (4.12),

$$\sigma_{ji} = \frac{\rho_0}{\det g} x_{i,A}x_{j,B} \sum_{\alpha=1}^{\lambda} \frac{\partial W}{\partial K_\alpha}\left(\frac{\partial K_\alpha}{\partial C_{AB}} + \frac{\partial K_\alpha}{\partial C_{BA}}\right) , \tag{5.10}$$

in the case when the material is compressible and

$$\sigma_{ji} = \rho_0 x_{i,A}x_{j,B} \sum_{\alpha=1}^{\lambda} \frac{\partial W}{\partial K_\alpha}\left(\frac{\partial K_\alpha}{\partial C_{AB}} + \frac{\partial K_\alpha}{\partial C_{BA}}\right) - p\delta_{ij} \tag{5.11}$$

in the case when it is incompressible.

For any specified material symmetry, the choice of the irreducible integrity basis $K_1,\ldots,K_\lambda$, in terms of which W is expressed, is not unique. It is, however, always possible, in the compressible case, to choose it so that three of the K's are I_1, I_2, I_3 defined by (5.2). In the incompressible case two of the K's can always be taken to be I_1 and I_2 . These facts may be of particular significance when one is concerned with the deformation of slightly anisotropic materials. It may then be possible to first solve the corresponding problem for an isotropic material and to calculate the effects on the solution of the small degree of anisotropy by using a perturbation procedure.

6. THE DETERMINATION OF THE STRAIN-ENERGY FUNCTION

In principle, the detailed dependence of the strain-energy function W on the strain invariants $K_1, K_2, \ldots, K_\lambda$, for a particular material, can be determined by carrying out appropriate simple experiments. From a practical point of view this involves considerable difficulty. It has, however, been done* in the case of vulcanized rubbers for which the material is isotropic and nearly enough incompressible so that the idealization of incompressibility can be made without significant error. We have seen that in this case W depends on only two invariants, I_1 and I_2, defined by (5.2).

The procedure adopted depends on the fact that in the incompressible isotropic case a number of simple problems, which can be readily modelled experimentally, can be solved [1,2,3] without making any further assumptions regarding W, beyond those implied by (5.7). Once the precise dependence of W on I_1 and I_2 has been determined, it can be used in the prediction of the results of other experiments. The range of such calculations has been greatly increased by the use of finite element computational techniques by Oden, Klosner and others (see, for example, [4]).

7. SMALL DEFORMATION THEORY

We shall now assume that the dependence of W on I_1, I_2 and I_3 is sufficiently smooth, so that it can be expressed as a power series in these quantities and hence as a power series in I_1-3, I_2-3, I_3-1, thus

$$W = \sum_{\alpha,\beta,\gamma} A_{\alpha\beta\gamma}(I_1-3)^\alpha(I_2-3)^\beta(I_3-1)^\gamma . \tag{7.1}$$

We note, from (5.2), that when the body is in its reference configuration, which we take as the undeformed state, $I_1=I_2=3$ and $I_3=1$. Also, we can, without loss of generality, take $A_{000} = 0$.

---

* This was first achieved by Rivlin and Saunders [11] in 1951. Since then analogous work on a wide variety of vulcanized rubbers has been carried out by many workers.

If the deformation is small, we can approximate (7.1) by a limited number of terms in the power-series. It is worth considering the physical implications of such approximations.

We note, as was shown in §2, that any deformation may locally be regarded as the resultant of a pure homogeneous deformation followed by a rigid rotation. In the pure homogeneous deformation a particle initially at $X+dX$ moves to $\xi+d\xi$, where

$$d\xi = s\, dX \tag{7.2}$$

and s is a symmetric tensor, such that (cf. equation (2.6)) $s^2=C$. Now, the principal values of s are the positive square roots of the principal values of C. We denote them by λ_1, λ_2 and λ_3. These are called the principal extension ratios for the deformation. In terms of them, I_1, I_2 and I_3 are given by

$$I_1 = \lambda_1^2 + \lambda_2^2 + \lambda_3^2, \quad I_2 = \lambda_2^2\lambda_3^2 + \lambda_3^2\lambda_1^2 + \lambda_1^2\lambda_2^2,$$
$$I_3 = \lambda_1^2\lambda_2^2\lambda_3^2. \tag{7.3}$$

We define the principal extensions e_α $(\alpha=1,2,3)$ by

$$e_\alpha = \lambda_\alpha - 1. \tag{7.4}$$

From (7.3) it follows that

$$\begin{aligned} I_1 - 3 &= 2E_1 + E_1^2 - 2E_2, \\ I_2 - 3 &= 4E_1 + 2E_1^2 + 2E_1E_2 - 6E_3 + O(e^4), \\ I_3 - 1 &= 2E_1 + E_1^2 + 2E_2 + 2E_3 + 2E_1E_2 + O(e^4), \end{aligned} \tag{7.5}$$

where

$$E_1 = e_1 + e_2 + e_3, \quad E_2 = e_2e_3 + e_3e_1 + e_1e_2, \quad E_3 = e_1e_2e_3, \tag{7.6}$$

and $O(e^4)$ denotes terms of degree higher than the third in the e's. We note that if the e's are sufficiently small compared with unity, the I's are, in general, of first order in the e's. Accordingly, if we wish to approximate W up to second-order in the e's, we have to retain in (7.1) nine terms; if we wish to approximate W up to third-order in the e's, we have to retain in (7.1) sixteen terms. Following Murnaghan [12], we can approximate (7.1) with fewer terms by replacing I_1-3, I_2-3, I_3-1 by three new invariants J_1, J_2, J_3 defined by

$$J_1 = I_1-3, \quad J_2 = I_2-2I_1+3, \quad J_3 = I_3-I_2+I_1-1 \ . \tag{7.7}$$

From (7.7) and (7.5), we obtain

$$\begin{aligned} J_1 &= 2E_1 + E_1^2 - 2E_2 \ , \\ J_2 &= 4E_2 + 2E_1E_2 - 6E_3 + O(e^4) \ , \\ J_3 &= 8E_3 + O(e^4) \ . \end{aligned} \tag{7.8}$$

With (7.7), the expression (7.1) for W may be expressed as

$$\rho_0 W = \Sigma \, B_{\alpha\beta\gamma} \, J_1^\alpha J_2^\beta J_3^\gamma \ , \qquad B_{000} = 0 \ . \tag{7.9}$$

If we wish to retain in (7.9) terms of first and second orders in the e's, we have, with an evident change of notation,

$$\rho_0 W = a_0 J_1 + a_1 J_2 + a_2 J_1^2 \ . \tag{7.10}$$

If we wish to retain terms of orders up to the third in the e's, we have

$$\rho_0 W = a_0 J_1 + a_1 J_2 + a_2 J_1^2 + a_3 J_1 J_2 + a_4 J_1^3 + a_5 J_3 \ . \tag{7.11}$$

In (7.10) and (7.11) the a's are, of course, physical constants for the material.

It can easily be shown that if we make the assumption that the stress is zero in the reference configuration, this implies that $B_{100} = 0$ in (7.9) and correspondingly that $a_0 = 0$ in (7.10) and (7.11). If this assumption is not made, then the Cauchy stress in the reference configuration is a hydrostatic tension of amount $2\rho_0 a_0$.

In deriving equations (7.10) and (7.11) from (7.9), the only assumption made is that the principal extensions are sufficiently small. No limitation is placed on the magnitude of the rigid rotation. If we also make the assumption that the rotation is small, then the displacement gradients are small. We introduce the notation $d = \| d_{AB} \| = \| u_{A,B} \|$. Then neglecting terms of higher degree than the third in d , we obtain

$$\begin{aligned} J_1 &= 2 \operatorname{tr} d + \operatorname{tr} dd^T \ , \\ J_2 &= 2(\operatorname{tr} d)^2 - \operatorname{tr} d^2 - \operatorname{tr} dd^T + 2 \operatorname{tr} d \operatorname{tr} dd^T \\ &\qquad - 2 \operatorname{tr} d^2 d^T \ , \\ J_3 &= 8 \det d \ . \end{aligned} \tag{7.12}$$

If we introduce these expressions for J_1, J_2, J_3 into (7.10) and neglect terms of third and higher degrees in d, we obtain

$$\rho_0 W = 2a_0 \operatorname{tr} d + (a_0 - a_1)\operatorname{tr} dd^T + 2(a_1 + 2a_2)(\operatorname{tr} d)^2 - a_1 \operatorname{tr} d^2 . \quad (7.13)$$

If we take $a_0=0$ in (7.13) corresponding to the assumption that the stress is zero in the reference state, equation (7.13) becomes

$$\rho_0 W = 2(a_1 + 2a_2)(\operatorname{tr} e)^2 - 2a_1 \operatorname{tr} e^2 , \quad (7.14)$$

where e is the classical infinitesimal strain tensor defined by

$$e = \frac{1}{2}(d + d^T) . \quad (7.15)$$

We note that if we write

$$\mu = -2a_1 \quad \text{and} \quad \lambda = 4(a_1 + 2a_2) \quad (7.16)$$

in (7.14), we obtain the usual formula for the strain-energy function in classical elasticity theory, λ and μ being the Lamé constants.

In the same way, we can introduce the expressions (7.12) into (7.11), with $a_0=0$, and neglect terms of higher degree than the third in d to obtain the next approximation to the strain energy function.

8. INITIAL STRESS THEORY

We now suppose that a body of elastic material is subjected to an initial finite deformation in which a particle initially at X moves to x and that we superpose on this deformation an infinitesimal deformation in which the particle moves to $x+\tilde{u}$. Let $\tilde{\Pi}$ denote the change in the Piola-Kirchhoff stress associated with this superposed infinitesimal deformation. Then, from (4.3), it follows that

$$\tilde{\Pi}_{Ai} = \rho_0 \frac{\partial^2 W}{\partial x_{i,A} \partial x_{j,B}} \tilde{u}_{j,B} . \quad (8.1)$$

It is easily seen (cf. (3.1) and (3.2)) that $\tilde{\Pi}_{Ai}$ must satisfy the equations of motion

$$\tilde{\Pi}_{Ai,A} + \rho_0 \tilde{\phi}_i = \rho_0 \ddot{\tilde{u}}_i \quad (8.2)$$

and the boundary conditions

$$\tilde{F}_i = \tilde{\Pi}_{Ai} N_A , \tag{8.3}$$

where $\tilde{\phi}$ is the additional body force per unit mass and $\tilde{F}$ is the additional surface force per unit area measured in the undeformed state.

In the case when the material considered possesses some symmetry, the appropriate form for W can, of course, be introduced into (8.1)

Using the equations so derived, a wide variety of problems have been solved of the so-called *initial stress* type. In these, the underlying finite deformation $X \to x$ is usually taken to be a relatively simple one - pure homogeneous or possibly, in the case when the material is incompressible, simple torsion. However, the superposed infinitesimal deformation may be considerably more complicated in character. As examples of problems of this type which have been solved we may mention the following:

(i) the effect of simple extension of a rod of circular [4], or more general [13,14] cross-section on the torsional modulus of the rod for small superposed torsions;

(ii) the effect of initial pure homogeneous deformation on the indentation hardness of a half-space, in the case when the principal extension ratios in the bounding surface of the half-space are equal [15];

(iii) the effect of initial pure homogeneous deformation on the velocity of propagation of plane sinusoidal waves of infinitesimal amplitude [16].

The initial stress theory has particular interest in connection with the study of stability. It has been applied (see, for example, [17] and references there given) to the determination of the bifurcation condition for edge loading of rods, strips and plates of elastic material.

9. MATERIALS WITH MEMORY

We again describe the deformation in a body in the manner adopted in §2, but we are now concerned with materials for which the stress measured at time t depends not only on the instantaneous value of the deformation gradient tensor at time t, at the particle considered, but also on the previous history of the deformation gradient tensor at this particle, from the infinite past up to and including the instant t. In mathematical terms this can be expressed by the assumption that the Piola-Kirchhoff stress, or the Cauchy stress tensor, at time t, is a functional of the history $g(\tau)$ with support $\tau = (-\infty, t]$, thus

$$\sigma = F[g(\tau)] . \tag{9.1}$$

We can now parallel the argument which led us to conclude,

for an elastic material, that the strain-energy at time t must depend on the value of the deformation gradient tensor at time t through the Cauchy strain tensor. In this case, we recognize [5,7,8] that if we superpose on the assumed deformation a time-dependent rigid rotation, the Cauchy stress tensor at time t is rotated by the amount of this rotation at time t . Let $r(\tau)$ be an arbitrary time-dependent rigid rotation and let

$$\bar{g}(\tau) = r(\tau)g(\tau) \; . \tag{9.2}$$

Then, the Cauchy stress $\bar{\sigma} = \bar{\sigma}(t)$ associated with the deformation gradient history $\bar{g}(\tau)$ is given by

$$\bar{\sigma} = F[\bar{g}(\tau)] = F[r(\tau)g(\tau)] \; . \tag{9.3}$$

The effect of a superposed rigid rotation on the stress can be expressed by the relation

$$\sigma = r^{T}\bar{\sigma}r \; , \tag{9.4}$$

where $r = r(t)$. From (9.1), (9.2) and (9.4), we obtain [7]

$$F[g(\tau)] = r^{T}F[r(\tau)g(\tau)]r \; . \tag{9.5}$$

Since this relation must be valid for all rigid rotations, it must, in particular, be valid for (cf. §4)

$$r(\tau) = \{C(\tau)\}^{-\frac{1}{2}}g^{T}(\tau) \; . \tag{9.6}$$

Introducing (9.6) into (9.5), we obtain

$$\sigma = F[g(\tau)] = r^{T}F[\{C(\tau)\}^{\frac{1}{2}}]r \; , \tag{9.7}$$

where

$$r = r(t) = C^{-\frac{1}{2}}g^{T} \; . \tag{9.8}$$

Thus, σ must be expressible in the form [8]

$$\sigma = r^{T}F[C(\tau)]r \; , \tag{9.9}$$

with r given by (9.8). Introducing (9.8) into (9.9), we obtain an alternative form [7] for σ

$$\sigma = gF[C(\tau)]g^{T} \; . \tag{9.10}$$

We note that the forms (9.7) and (9.10) are such that they satisfy (9.5) for arbitrary rigid rotations r .

We shall now consider somewhat more critically the constitutive assumption (9.1) which leads to (9.9) and (9.10). At first sight this assumption, expressing at it does the physical consideration that the stress at time t depends on the deformation gradients up to and including time t, appears to be hardly more than an expression of the law of causality. In examining it further, we note firstly that the constitutive assumption $(4.1)_2$ which is made for an elastic solid is a special case of (9.1). For, if $g(\tau)$ is known for $\tau = (-\infty,t]$, then $g = g(t)$ is known; i.e. g is a functional of $g(\tau)$ with support $(-\infty,t]$. Thus, the relation

$$\sigma = \sigma(g) , \tag{9.11}$$

the Cauchy stress tensor at time t is an ordinary function of the deformation gradient tensor at time t, is a special case of (9.1). With (3.7), we see that $(4.1)_2$ is a special case of (9.1).

Now, let us examine the status of constitutive assumptions of the form

$$\sigma = \sigma(g,\dot{g}) , \tag{9.12}$$

where $\dot{g} = dg(\tau)/d\tau|_{\tau=t}$, special cases of which provide the constitutive equations for a Kelvin solid and a Newtonian fluid. We note that $\dot{g}$ is not necessarily a functional of $g(\tau)$ with support $(-\infty,t]$. To see this, we have only to consider two deformation gradient histories $g(\tau)$. In one of these $g(\tau) = \delta$ for all τ. In the other, $g(\tau) = \delta$ for $\tau = (-\infty,t]$ and $g(\tau)-\delta = \kappa(\tau-t)$, where κ is a constant tensor, for $\tau \geq t$; i.e. $dg(\tau)/d\tau$ changes discontinuously at time t. According to (9.1), the stress at time t is the same for both deformation gradient histories. For a Kelvin solid or a Newtonian fluid, this is not the case. Of course, one might argue that in no physical situation can $\dot{g}$ change strictly discontinuously. However, if $\dot{g}$ changes very rapidly at time t, though not discontinuously, the expression for $\dot{g}$ as a functional of $g(\tau)$ with support $(-\infty,t]$ would be difficult to handle. Accordingly, if we wish our constitutive assumption to include relations of the form (9.12) as special cases, we may write it in a variety of forms [18,19,20]:

$$\begin{aligned}
&\sigma = F[g(\tau);\dot{g}] , && \tau = (-\infty,t] \\
&\sigma = F[g(\tau);g,\dot{g}] , && \tau = (-\infty,t] \\
&\sigma = F[dg(\tau)/d\tau] , && \tau = (-\infty,t] \\
&\sigma = \underset{\varepsilon\to 0}{\mathrm{Lt}}\, F[g(\tau)] , && \tau = (-\infty,t+\varepsilon) .
\end{aligned} \tag{9.13}$$

In the last case, we recognize that $\dot{g}$ is a functional of $g(\tau)$

with support $\tau = (-\infty, t+\varepsilon)$, thus

$$\dot{g} = -\int_{t-\varepsilon}^{t+\varepsilon} \delta'(\tau - t) g(\tau) d\tau \ . \tag{9.14}$$

The restrictions, resulting from the consideration of the effect on the Cauchy stress of a superposed rigid rotation, can be introduced, in each of the cases (9.13), in a manner similar to that used in deriving (9.9) and (9.10). We find that σ must be expressible in the forms

$$\begin{aligned} \sigma &= r^T F[C(\tau);\dot{C}]r \ , \quad \tau = (-\infty,t] \\ \sigma &= r^T F[C(\tau);C,\dot{C}]r \ , \quad \tau = (-\infty,t] \\ \sigma &= r^T F[dC(\tau)/d\tau]r \ , \quad \tau = (-\infty,t] \\ \sigma &= r^T \operatorname*{Lt}_{\varepsilon\to 0} F[C(\tau)]r \ , \quad \tau = (-\infty,t+\varepsilon) \ . \end{aligned} \tag{9.15}$$

respectively, where $C = C(t)$ and $\dot{C} = dC(\tau)/d\tau|_{\tau=t}$, and r is given by (9.8). Introducing (9.8) into (9.15), we obtain, paralleling (9.10), the alternative forms

$$\begin{aligned} \sigma &= gF[C(\tau);\dot{C}]g^T \ , \quad \tau = (-\infty,t] \\ \sigma &= gF[C(\tau);C,\dot{C}]g^T \ , \quad \tau = (-\infty,t] \\ \sigma &= gF[dC(\tau)/d\tau]g^T \ , \quad \tau = (-\infty,t] \\ \sigma &= g \operatorname*{Lt}_{\varepsilon\to 0} F[C(\tau)]g^T \ , \quad \tau = (-\infty,t+\varepsilon) \end{aligned} \tag{9.16}$$

respectively. We note that if we superpose on the assumed deformation an arbitrary rigid rotation, the functional F remains unchanged if σ is expressed in the form (9.15), while it changes if the form (9.16) is adopted.

Corresponding expressions to (9.15) and (9.16) can, of course, be obtained by using the relation (3.7).

In general, each of the functionals in this section may depend on the time t , as well as on the indicated arguments. For hereditary materials, the stress at time t , resulting from a deformation gradient history $g(\tau)$, is the same as the stress at time $t+\bar{t}$, say, resulting from a deformation gradient history $g(\tau+\bar{t})$, where $\bar{t}$ is arbitrary. It follows that (9.1) may be expressed in the form

$$\sigma = F[g(t-s)] \ , \tag{9.17}$$

where $s = [0,\infty)$ and it is understood that the form of the functional F is independent of t . Accordingly, the relation (9.9) may be expressed in the form

$$\sigma = r^T F[C(t-s)]r \quad (9.18)$$

where, again, $s = [0,\infty)$ and the form of the functional F is independent of t . Analogous changes may be made in the constitutive assumptions (9.13) and the constitutive equations (9.15) and (9.16) resulting from them, to express the fact that the materials considered are of the hereditary type.

For an incompressible material, the constitutive assumption (9.1) must be replaced by

$$\sigma = F[g(\tau)] - p\delta \ , \quad (9.19)$$

where p is undetermined if the history $g(\tau)$ is specified. Analogous changes must be made in the constitutive assumptions (9.13) and in the constitutive equations (9.10), (9.11), (9.15) and (9.16) derived from (9.1) and (9.13).

10. MATERIAL SYMMETRY IN MATERIALS WITH MEMORY

If the material possesses some symmetry in its reference configuration, described by the group of transformations $\{S\}$, then the tensor functional F in (9.9) must satisfy the relation

$$F[\bar{C}(\tau)] = S\ F[C(\tau)]S^T \ , \quad (10.1)$$

where

$$\bar{C}(\tau) = S\ C(\tau)S^T \ , \quad (10.2)$$

for all transformations in the group $\{S\}$. This condition parallels the condition (5.1) which is imposed on the strain-energy for an elastic material by material symmetry. As in that case the next step is to render the implicit condition (10.1) explicit. This can, in fact, be done, without further assumptions, for isotropic materials, for materials having transverse isotropy, and for any of the crystalline symmetries.

11. INTEGRAL REPRESENTATION OF FUNCTIONALS

So long as the tensor functionals F in (9.15) and (9.16) are left in the general forms given there, even as modified by any applicable symmetry condition of the form (9.1), rather little progress can be made. Accordingly, more explicit expressions must be obtained for them. To do so, some sacrifice of generality has to be made. This usually takes the form of some assumption regarding the smoothness of dependence of F on its arguments. A wide variety of assumptions of this type can be made and each one of them entails different implications with respect to the physical behavior of the material described and the class of allowable deformation histories.

In the present paper we will discuss the implications of two such assumptions. In doing so, it is more convenient to regard F as a tensor functional of the history of the reduced Cauchy strain $E(\tau)$, defined by

$$E(\tau) = C(\tau) - \delta , \tag{11.1}$$

rather than of the history of the Cauchy strain. The relation (9.9) can then be written in the form

$$\sigma = r^T F[E(\tau)]r . \tag{11.2}$$

We note that $E(\tau) = 0$ when the material is undeformed.

In order to introduce a continuity assumption on F, it is necessary to define the "distance" between two tensor functions $E(\tau)$, i.e. to define a metric for the space of tensor functions $E(\tau)$. We shall define the "distance" $|E_1(\tau)-E_2(\tau)|$ between two tensor functions $E_1(\tau)$ and $E_2(\tau)$ by

$$|E_1(\tau)-E_2(\tau)| = \sup\{\mathrm{tr}[E_1(\tau)-E_2(\tau)]^2\}^{\frac{1}{2}} . \tag{11.3}$$

Let $F_1 = F[E_1(\tau)]$ and $F_2 = F[E_2(\tau)]$. We shall define the distance $|F_1-F_2|$ between F_1 and F_2 by

$$|F_1-F_2| = \{\mathrm{tr}[F_1-F_2]^2\}^{\frac{1}{2}} . \tag{11.4}$$

We now assume that F is a uniformly continuous functional of $E(\tau)$, in the sense of the metrics defined by (11.3) and (11.4), over some space of functions $E(\tau)$. This means that for any $\varepsilon>0$ there exists a quantity δ, which depends on ε, such that

$$\text{if} \quad |F_1-F_2| < \varepsilon , \quad \text{then} \quad |E_1(\tau)-E_2(\tau)| < \delta . \tag{11.5}$$

We now consider histories for which $E(\tau) = 0$ for $\tau<0$ and for which $E(\tau)$ can be expressed in the form of a uniformly convergent cosine Fourier series with half-period t in the interval $[0,t]$, where t is some fixed time, thus

$$E(\tau) = \sum_{\alpha=0}^{\infty} E_\alpha \cos \alpha\pi\tau/t , \tag{11.6}$$

where

$$E_\alpha = 2\int_0^t E(\tau)\cos\frac{\alpha\pi\tau}{t}\,d\tau\ , \qquad (\alpha=1,2,\ldots)$$
$$E_0 = \int_0^t E(\tau)d\tau\ . \tag{11.7}$$

We now define $E^{(\mu)}(\tau)$ by

$$E^{(\mu)}(\tau) = \sum_{\alpha=0}^{\mu} E_\alpha \cos\alpha\pi\tau/t\ , \tag{11.8}$$

where μ is some fixed number, and write

$$E(\tau) = E^{(\mu)}(\tau) + R^{(\mu)}(\tau)\ . \tag{11.9}$$

We limit our discussion to histories of the reduced Cauchy strain for which

$$\mathrm{tr}\{E_\alpha\}^2 < B_\alpha\ , \text{ say, } (\alpha=0,1,\ldots,\mu)\ , \tag{11.10}$$

where B_α are specified quantities. This condition ensures that the space of tensor functions $E^{(\mu)}(\tau)$ is compact. It then follows from the Stone-Weierstrass theorem that $F[E^{(\mu)}(\tau)]$ can be uniformly approximated with any desired accuracy, ε say, by a polynomial in E_α $(\alpha=0,1,\ldots,\mu)$. Let us denote such a polynomial by $P(E_\alpha;\varepsilon)$. Then,

$$|F[E^{(\mu)}(\tau)] - P(E_\alpha;\varepsilon)| < \varepsilon\ . \tag{11.11}$$

Now, suppose that

$$|F[E(\tau)] - F[E^{(\mu)}(\tau)]| < \varepsilon\ . \tag{11.12}$$

In view of the continuity of F, expressed by (11.5), it follows that

$$|E(\tau)-E^{(\mu)}(\tau)| < \delta\ . \tag{11.13}$$

Since (11.6) is uniformly convergent, this can always be achieved by the functions defined by (11.6) and (11.8), provided that a sufficiently large value of μ is taken. From (11.11) and (11.12), it follows that

$$|F[E(\tau)] - P(E_\alpha;\varepsilon)| < 2\varepsilon\ ; \tag{11.14}$$

i.e. $F[E(\tau)]$ can be uniformly approximated, with any desired accuracy, by a tensor polynomial in E_α $(\alpha=0,1,\ldots,\mu)$ if μ is large enough. Since any monomial in the E_α's can be expressed as a multiple integral of multiplicity equal to the degree of the monomial, we conclude that we can approximate F uniformly, with any desired accuracy, 2ε say, by the sum of a series of multiple integrals thus

$$F[E(\tau)] = \sum_{\lambda=1}^{\nu} F_\lambda , \tag{11.15}$$

where $F_\lambda = \| F_{ij}^{(\lambda)} \|$ is given by

$$F_{ij}^{(\lambda)} = \int_0^t \cdots \int f_{iji_1j_1\ldots i_\lambda j_\lambda} (\tau_1/t,\ldots,\tau_\lambda/t;\varepsilon)$$

$$E_{i_1j_1}(\tau_1)\ldots E_{i_\lambda j_\lambda}(\tau_\lambda)d\tau_1\ldots d\tau_\lambda \tag{11.16}$$

and the kernels $f_{ij\ldots i_\lambda j_\lambda}$ are continuous functions of $\tau_1,\ldots,\tau_\lambda$. In (11.15) we have taken $F_0=0$ in order to satisfy the condition that $F=0$ when $E(\tau) = 0$. If t is allowed to take a range of values, then it occurs as an argument of the kernels in (11.16).

It is, of course, evident that precisely the same considerations can be applied, taking any of equations (9.15) as our starting point. For example, if $(9.15)_1$ is taken as a starting point, we arrive at an expression for σ of the form

$$\sigma = r^T F[E(\tau);\dot{C}]r , \tag{11.17}$$

with

$$F[E(\tau);\dot{C}] = F_0 + \sum_{\lambda=1}^{\nu} F_\lambda , \tag{11.18}$$

where F_λ is given by (11.16), the kernels now being functions of $\dot{C}$, and F_0 is a tensor function of $\dot{C}$.

It will, of course, be recognized that the set of conditions which have been imposed here on the functional F and on the space of functions $E(\tau)$, in order to allow us to approximate it by an expression of the form given by (11.15) and (11.16), are only one of many such sets of conditions.

For hereditary materials, the expression for F_λ in (11.15) can be modified slightly from the form given in (11.16). We note that the kernels in (11.16) are, in general, functions of the indicated arguments and of t. Since, for a hereditary material, the constitutive assumption (9.1) is replaced by one of the form (9.18), where the form of the functional F is independent of t , the expression (11.16) can be replaced by one of the form

$$F_{ij}^{(\lambda)} = \int_t^0 \cdots \int f_{iji_1j_1\ldots i_\lambda j_\lambda}(s_1,\ldots,s_\lambda;\varepsilon)$$

$$E_{i_1j_1}(t-s_1)\ldots E_{i_\lambda j_\lambda}(t-s_\lambda)ds_1\ldots ds_\lambda \ , \qquad (11.19)$$

where the kernel is independent of t and is a continuous function of the indicated arguments. This may, of course, be rewritten as

$$F_{ij}^{(\lambda)} = \int_0^t \cdots \int f_{iji_1j_1\ldots i_\lambda j_\lambda}(t-\tau_1,\ldots,t-\tau_\lambda;\varepsilon)$$

$$E_{i_1j_1}(\tau_1)\ldots E_{i_\lambda j_\lambda}(\tau_\lambda)d\tau_1\ldots d\tau_\lambda \ . \qquad (11.20)$$

12. REPRESENTATION OF FUNCTIONALS BY DERIVATIVES

Instead of assuming, as we have done in the previous section, that $E(\tau)$ is expressible as a uniformly convergent Fourier series, we now assume that it is expressible as a uniformly convergent Taylor series about time t , thus

$$E(\tau) = \sum_{\alpha=0}^{\infty} \frac{1}{\alpha!}(\tau-t)^\alpha E_\alpha \ , \qquad (12.1)$$

with the notation

$$E_0 = E = E(t) \ , \quad E_\alpha = \left.\frac{d^\alpha E(\tau)}{d\tau^\alpha}\right|_{\tau=t} \ . \qquad (12.2)$$

We again adopt the supremum metric, defined by (11.3), as the measure of the distance between two functions and we again define the distance between two values of F by (11.4). We also assume that F is a uniformly continuous functional of $E(\tau)$ in the sense of these metrics, so that (11.5) is applicable.

Paralleling the discussion of the previous section, we now define $E^{(\mu)}(\tau)$ by

$$E^{(\mu)}(\tau) = \sum_{\alpha=0}^{\mu} \frac{1}{\alpha!}(\tau-t)^\alpha E_\alpha \ , \qquad (12.3)$$

where μ is some fixed number, and write

$$E(\tau) = E^{(\mu)}(\tau) + R^{(\mu)}(\tau) \ . \qquad (12.4)$$

We limit our discussion to histories of the reduced Cauchy strain for which

$$\mathrm{tr}\{E_\alpha\}^2 < B_\alpha \;,\quad \text{say,}\quad (\alpha=0,1,\ldots,\mu) \;, \tag{12.5}$$

where B_α are specified quantities. This condition ensures that the space of tensor functions $E^{(\mu)}(\tau)$ is compact. It then follows from the Stone-Weierstrass theorem that $\mathcal{F}[E^{(\mu)}(\tau)]$ can be uniformly approximated with any desired accuracy, ε say, by a polynomial in E_α $(\alpha=0,1,\ldots,\mu)$. Following an argument analogous to that employed in the previous section, we can then show that provided μ is sufficiently large, we can uniformly approximate $\mathcal{F}[E(\tau)]$ with accuracy 2ε by a polynomial in E_α $(\alpha=0,1,\ldots,\mu)$, thus [6,7]

$$\mathcal{F}[E(\tau)] = F(E_\alpha) \;. \tag{12.6}$$

In general, the form of F will depend on t, but this dependence disappears if the material is of the hereditary type.

We note that we arrive at a similar form for $\mathcal{F}$ by taking as our starting point either of the constitutive equations $(9.15)_{1,2}$. We also arrive at a similar form if we take, as our starting point, the constitutive equation $(9.15)_4$ and assume that $E(\tau)$ is expressible as a uniformly convergent Taylor series about t, in the interval $(0,t+\varepsilon)$. Again, starting with $(9.15)_3$, we can arrive at a form similar to (9.6), with the argument E_0 absent, by assuming that $dE(\tau)/d\tau$ is expressible as a uniformly convergent Taylor series about time t.

13. SPECIAL CASES OF THE REPRESENTATIONS

While the approximate representation for $\mathcal{F}$ given by (11.15) and (11.16) may be formally valid, with any desired accuracy, under the conditions stated in §11, its practical value is rather limited unless the number of multiple integrals in the representation, i.e. ν, is small. Otherwise, the experimental determination of the kernels in (11.16) would be impossible to achieve. In general, this is only the case if the space of functions over which the representation is valid is one for which rather few terms in the Fourier series (11.6) provides a good approximation to the strain history considered, if the bounds on the coefficients is sufficiently small, and if the range of values of t is small.

Similarly, the representation given by (12.6) will also be of rather limited value unless μ is small and the degree of the tensor polynomial in (12.6) is small. In general this requires that the strain history vary sufficiently slowly with time and that the strain be sufficiently small.

It should be noted that, in both of the cases discussed, representations of the type derived can approximate F with any desired *absolute* accuracy, but not with any desired percentage accuracy.

Notwithstanding the above comments, no rule can be given in physical terms which will render valid, for all materials, an approximation for F either by few terms in the integral representation (11.15) or by a polynomial of the type (12.6) of low degree.

14. FADING MEMORY

For many materials with memory, the stress measured at time t, say, "forgets" that part of the deformation history which lies in the past prior to t. Such materials are said to exhibit *fading memory*. For example, it is evident that if the kernels in the multiple integrals $F_{ij}^{(\lambda)}$ given by (11.20) decrease as the arguments $t-\tau_1,\ldots,t-\tau_\lambda$ increase, then the corresponding material will exhibit fading memory.

We can give more formal expression to this type of material behavior in a variety of ways. One way in which this can be done is to consider two deformation gradient histories, for which the reduced Cauchy strains are $E(\tau)$ and $\bar{E}(\tau)$ respectively. We suppose that $E(\tau) = \bar{E}(\tau)$ in the interval $[t-s,t]$, where $s>0$ and $E(\tau) \neq \bar{E}(\tau)$ in part, or the whole of the interval $(-\infty,t-s)$. Let σ and $\bar{\sigma}$ be the Cauchy stresses at time t corresponding to the histories $E(\tau)$ and $\bar{E}(\tau)$ respectively. Then, if $\mathrm{tr}(\sigma-\bar{\sigma})^2$ tends uniformly to zero as $s\to\infty$, the material will exhibit fading memory.

If this is the case, then we may approximate the functional F, with any desired accuracy, by a functional with support $s = [0,\bar{s}]$ provided that $\bar{s}$ is made large enough. Provided that $t-\bar{s} > 0$ this enables us to replace the lower limit of integration by $t-\bar{s}$ in the integral representation provided by (11.15) with (11.20).

15. SOLUTION OF PROBLEMS IN NON-LINEAR VISCOELASTICITY

For the most part, the types of problem which have been found to be tractable in finite elasticity theory provide a guide to the types of problem which one may with benefit analyze on the basis of the various constitutive equations for non-linear viscoelastic materials which have been discussed. However, the much greater complexity of the constitutive equations in the case of viscoelastic materials and the large number of scalar functions which must be determined to characterize the material, even when continuity assumptions are introduced to validate the integral or differential representations, makes the experimental characterization of a particular material a forbidding and perhaps impossible task.

Nevertheless, the various types of solutions which were obtained in finite elasticity theory have their counterparts in the theory of non-linear viscoelastic materials. The status of the results may, however, be very different. Thus, although it is possible to solve certain problems having a high degree of geometrical symmetry, without making assumptions regarding the material beyond isotropy and incompressibility, they cannot be used to characterize the material with anything like the completeness of the corresponding results in finite elasticity theory. However, relations may be established between certain of these results, without such a complete characterization of the material being achieved.

ACKNOWLEDGEMENT

This paper was written with the support of a grant from the National Science Foundation to Lehigh University.

REFERENCES

1. R.S. Rivlin, Phil Trans. Roy. Soc. A 241, 379 (1948).
2. R.S. Rivlin, Proc. Roy. Soc. A 195, 463 (1949).
3. R.S. Rivlin, Phil. Trans. Roy. Soc. A 242, 173 (1949).
4. J.T. Oden, Finite Elements of Nonlinear Continua, McGraw-Hill, New York (1972).
5. J.G. Oldroyd, Proc. Roy. Soc. A 200, 523 (1950).
6. R.S. Rivlin and J.L. Ericksen, J. Rational Mech. Anal. 4, 323 (1955).
7. A.E. Green and R.S. Rivlin, Arch. Rational Mech. Anal. 1, 1 (1957).
8. W. Noll, Arch. Rational Mech. Anal. 2, 197 (1958).
9. J.L. Ericksen and R.S. Rivlin, J. Rational Mech. Anal. 3, 281 (1954).
10. G.F. Smith and R.S. Rivlin, Arch. Rational Mech. Anal. 1, 107 (1957).
11. R.S. Rivlin and D.W. Saunders, Phil. Trans. Roy. Soc. A, 243, 251 (1951).
12. F.D. Murnaghan, Amer. J. Math. 59, 235 (1937).
13. A.E. Green and R.T. Shield, Proc. Roy. Soc. A 202, 407 (1950).
14. A.N. Gent and R.S. Rivlin, Proc. Phys. Soc. B 65, 645 (1952).
15. A.E. Green, R.S. Rivlin and R.T. Shield, Proc. Roy. Soc. A 211, 128 (1952).
16. M. Hayes and R.S. Rivlin, Arch. Rational Mech. Anal. 8, 15 (1952).

17. K.N. Sawyers and R.S. Rivlin, Int. J. Solids Structures, 10, 483 (1974).
18. A.E. Green, R.S. Rivlin and A.J.M. Spencer, Arch. Rational Mech. Anal. 3, 82 (1959).
19. A.E. Green and R.S. Rivlin, Arch. Rational Mech. Anal. 4, 387 (1960).
20. R.S. Rivlin, "Materials with Memory" in Deformation and Fracture of High Polymers, ed. H.H. Kausch, J.A. Hassell and R.I. Jaffee, Plenum Press, New York (1973).

ANISOTROPIC ELASTIC AND PLASTIC MATERIALS

Tryfan G. Rogers

Department of Theoretical Mechanics,
University of Nottingham, England.

1. INTRODUCTION

The theory of the mechanical behaviour of anisotropic materials has been the subject of considerable interest and study for a great many years, particularly for those workers involved with natural structural materials such as wood or slate, and with layered materials, such as plywood or the earth.

Recently research in the subject has received a great impetus because of the introduction and widespread use of man-made fibre-reinforced materials. In these a weak, isotropic matrix with certain desirable properties such as lightness or ductility is reinforced in one or two particular directions by the introduction of very strong reinforcing cords or fibres in those directions. Macroscopically, these composite materials, when tested, will of course exhibit mechanical properties which are transversely isotropic if reinforced by one family of fibres, or orthotropic if reinforced in two directions (Figure 1). Furthermore, their response is not just anisotropic but is indeed highly anisotropic, so that isotropic theory would not provide even a rough approximation to their behaviour under most types of loading conditions. With these considerations in mind, a particularly simple continuum model has been proposed and developed for fibre-reinforced materials that can, at least to a good first approximation, give solutions to a number of significant boundary-value problems involving elastic, plastic or viscoelastic behaviour. Also the theory can treat large deformations as well as small ones, and the reinforcements can lie in curved lines as well as straight.

Thoft-Christensen (ed.), Continuum Mechanics Aspects of Geodynamics and Rock Fracture Mechanics, 177-200.

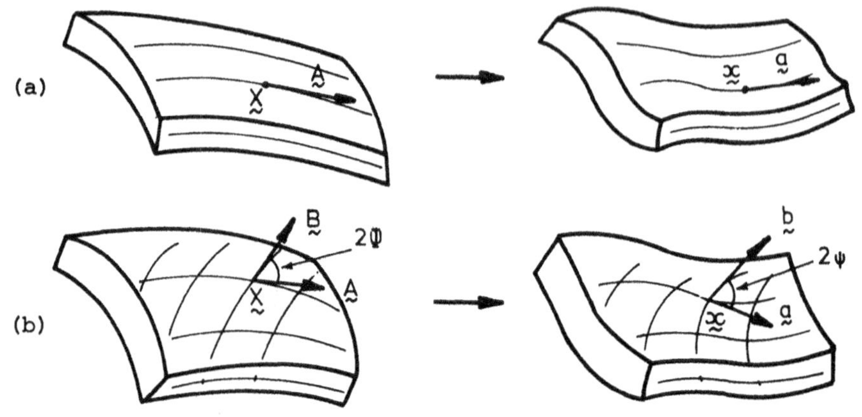

Fig. 1. Anisotropic materials exhibiting one or two preferred directions; (a) transverse isotropy, (b) orthotropy.

The success of such a theory suggests that it would be worthwhile to investigate other physical or engineering contexts in which it might be fruitfully applied. After all, the ultimate test of the validity of the appropriate constitutive equations is whether or nor they adequately describe the observed macroscopic behaviour of the material; the fact that its microstructure might include strong fibres then becomes irrelevant. The layered nature of many regions near the earth's surface indicates that this type of theory may be appropriate in some aspects of geophysics, particularly in situations of plane strain. The rock layers take the role of the fibres whilst the soil or softer layers respond like a weak matrix. All the phases may be isotropic, but for phenomena with characteristic length scales of several thicknesses the response is effectively anisotropic.

This paper basically falls into two halves, the first briefly reviewing the conventional theories for anisotropic solids with the remainder devoted to a description of the theory of strongly anisotropic materials. Section 2 gives the constitutive equations for anisotropic linear elasticity in a rather unconventional form which is particularly appropriate for materials with definite preferred directions. The relevant equations for finite elasticity are given in Section 3 and plasticity is dealt with in Section 4, with the formulation again influenced by possible application to materials exhibiting strength in particular directions. Section 5 introduces the foundations of the idealized theory of strongly anisotropic materials, and the very general results for plane deformations are obtained in Section 6. The final section discusses the extension of this theory to less idealized models and its application to "real" materials.

2. LINEAR ELASTICITY

The best-known constitutive equations for anisotropic materials are those associated with linearly elastic behaviour. Their most general form is

$$\sigma_{ij} = c_{ijk\ell}\varepsilon_{k\ell}, \qquad (i,j,k,\ell = 1,2,3) \tag{2.1}$$

where the σ_{ij} denote (Cauchy) stress components, ε_{ij} infinitesimal strain components with

$$\varepsilon_{ij} = \frac{1}{2}\left(\frac{\partial u_i}{\partial x_j} + \frac{\partial u_j}{\partial x_i}\right)$$

where $\underset{\sim}{u} = (u_1,u_2,u_3)$ is the displacement field, and the elastic constants $c_{ijk\ell}$ satisfy

$$c_{ijk\ell} = c_{jik\ell} = c_{ij\ell k} = c_{k\ell ij}. \tag{2.2}$$

In (2.1) and for the rest of this paper, we use the usual convention that a repeated suffix implies summation over the entire range of that suffix; this will be 1,2,3 unless otherwise stated.

The relations (2.2) imply that for the most general form of anisotropic behaviour the 81 elastic constants actually involve only 21 independent quantities. Any elastic symmetry further reduces the number and it is easily shown [1] that an orthotropic material has thirteen independent constants if the elastic symmetry is for reflection in only one plane and only nine if the symmetry is for reflection in two perpendicular planes. For transversely isotropic materials the number falls to five [2]. The constitutive equations are usually found to be written for convenience in terms of a coordinate system which coincides with the planes of symmetry for orthotropic behaviour, or with one of the axes coinciding with the preferred direction for transversely isotropic behaviour. However, this choice of "local" axes may have to change from point to point in some problems, particularly those involving a curved preferred direction; furthermore it is often very useful to have the preferred directions specifically incorporated in the formulation but without requiring a particular choice of axes of reference.

For <u>transverse isotropy</u> with $\underset{\sim}{a} \equiv (a_1,a_2,a_3)$ denoting the single axis of symmetry or preferred direction, dependent on position, a convenient form of the constitutive equations is [3]

$$\begin{aligned}\sigma_{ij} &= \lambda\varepsilon_{kk}\delta_{ij} + 2\mu_T\varepsilon_{ij} + \alpha(a_k a_m \varepsilon_{km}\delta_{ij} + a_i a_j \varepsilon_{kk}) \\ &\quad + 2(\mu_L - \mu_T)(a_i a_k \varepsilon_{kj} + a_j a_k \varepsilon_{ki}) + \beta a_i a_j a_k a_m \varepsilon_{km}\end{aligned} \tag{2.3}$$

in indicial notation, with components referred to cartesian axes, or

$$\begin{aligned}\underset{\sim}{\sigma} &= \lambda \underset{\sim}{I}\,\mathrm{tr}\,\underset{\sim}{\varepsilon} + 2\mu_T \underset{\sim}{\varepsilon} + \alpha\{\mathrm{tr}\,(\underset{\sim}{a}\,\underset{\sim}{a}\,\underset{\sim}{\varepsilon})\underset{\sim}{I} + \underset{\sim}{a}\,\underset{\sim}{a}\,\mathrm{tr}\,\underset{\sim}{\varepsilon}\} \\ &\quad + 2(\mu_L - \mu_T)(\underset{\sim}{a}\,\underset{\sim}{a}\,\underset{\sim}{\varepsilon} + \underset{\sim}{\varepsilon}\,\underset{\sim}{a}\,\underset{\sim}{a}) + \beta \underset{\sim}{a}\,\underset{\sim}{a}\,\mathrm{tr}\,(\underset{\sim}{a}\,\underset{\sim}{a}\,\underset{\sim}{\varepsilon})\end{aligned}$$

in direct notation, with $\underset{\sim}{I}$ denoting the unit tensor and $\mathrm{tr}\,\underset{\sim}{\varepsilon}$ denoting the trace of the matrix $\underset{\sim}{\varepsilon}$:

$$\mathrm{tr}\,\underset{\sim}{\varepsilon} \equiv \varepsilon_{11} + \varepsilon_{22} + \varepsilon_{33} = \varepsilon_{kk}.$$

In (2.3) μ_L and μ_T represent the shear moduli for shear respectively along and transverse to the preferred direction $\underset{\sim}{a}$. Note that by putting $\underset{\sim}{a} = (1,0,0)$ the equation reduces to the conventional form:

$$\begin{aligned}\sigma_{11} &= c_{11}\varepsilon_{11} + c_{12}\varepsilon_{22} + c_{13}\varepsilon_{33}, \\ \sigma_{22} &= c_{12}\varepsilon_{11} + c_{22}\varepsilon_{22} + c_{23}\varepsilon_{33}, \\ \sigma_{33} &= c_{13}\varepsilon_{11} + c_{23}\varepsilon_{22} + c_{33}\varepsilon_{33}, \\ \sigma_{23} &= 2c_{44}\varepsilon_{23}, \quad \sigma_{13} = 2c_{55}\varepsilon_{13}, \quad \sigma_{12} = 2c_{66}\varepsilon_{12}\end{aligned}$$

where the elastic moduli c_{rs} $(r,s = 1,2,\ldots,6)$ are related to $\lambda, \mu_L, \mu_T, \alpha$ and β through

$$\begin{aligned}c_{11} &= \lambda + 2\alpha + 4\mu_L - 2\mu_T + \beta, \quad c_{12} = c_{13} = \lambda + \alpha, \quad c_{23} = \lambda \\ c_{22} &= c_{33} = \lambda + 2\mu_T, \quad c_{44} = \mu_L, \quad c_{55} = c_{66} = \mu_T.\end{aligned}$$

Alternatively the strains may be expressed in terms of the stresses:

$$\begin{aligned}\varepsilon_{11} &= (\sigma_{11} - \nu\sigma_{22} - \nu\sigma_{33})/E \\ \varepsilon_{22} &= -\nu\sigma_{11}/E + (\sigma_{22} - \nu'\sigma_{33})/E' \\ \varepsilon_{33} &= -\nu\sigma_{11}/E - (\nu\sigma_{22} - \sigma_{33})/E' \\ \varepsilon_{23} &= \tfrac{1}{2}\sigma_{23}/\mu_L, \quad \varepsilon_{13} = \tfrac{1}{2}\sigma_{13}/\mu_T, \quad \varepsilon_{12} = \tfrac{1}{2}\sigma_{12}/\mu_T\end{aligned} \tag{2.4}$$

where E, ν and E', ν' are the Young's moduli and Poisson ratios in the longitudinal and transverse directions respectively, with

$$\nu' = \frac{E'}{2\mu_L} - 1. \tag{2.5}$$

The stress-strain relations for isotropic behaviour are given by $\mu_L = \mu_T$, $\alpha = \beta = 0$ or equivalently $E = E'$, $\nu = \nu'$.

If the material has two preferred directions $\underset{\sim}{a}$ and $\underset{\sim}{b}$ its behaviour is locally orthotropic. From results in invariant theory (a general account is given by Spencer [4]), and retaining only the linear terms in strain, we can deduce

$$\begin{aligned}\sigma_{ij} = {} & \lambda\varepsilon_{kk}\delta_{ij} + 2\mu\varepsilon_{ij} + \alpha_1(a_i a_k\varepsilon_{kj} + a_j a_k\varepsilon_{ki}) + \alpha_2(b_i b_k\varepsilon_{kj} + b_j b_k\varepsilon_{ki}) \\ & + \{\alpha_3 a_i a_j + \alpha_4 b_i b_j + \alpha_5(a_i b_j + a_j b_i)\}\, a_r b_s\varepsilon_{rs} \\ & + \alpha_6(a_r a_s\varepsilon_{rs}\delta_{ij} + a_i a_j\varepsilon_{kk}) + \alpha_7(b_r b_s\varepsilon_{rs}\delta_{ij} + b_i b_j\varepsilon_{kk}) \\ & + \alpha_8\{(a_r b_s + a_s b_r)\varepsilon_{rs}\delta_{ij} + (a_i b_j + a_j b_i)\varepsilon_{kk}\} \\ & + \alpha_9 a_i a_j a_r a_s\varepsilon_{rs} + \alpha_{10} b_i b_j b_r b_s\varepsilon_{rs} \\ & + \alpha_{11}(a_i a_j b_r b_s + b_i b_j a_r a_s)\varepsilon_{rs},\end{aligned} \tag{2.6}$$

with thirteen independent constants $\lambda, \mu, \alpha_1, \ldots, \alpha_{11}$. If the two directions $\underset{\sim}{a}$ and $\underset{\sim}{b}$ are mechanically equivalent,

$$\alpha_1 = \alpha_2, \quad \alpha_3 = \alpha_4, \quad \alpha_6 = \alpha_7, \quad \alpha_9 = \alpha_{10}$$

and the number of independent constants reduces to nine. Putting $\underset{\sim}{a} = (1,0,0)$ and $\underset{\sim}{b} = (0,1,0)$ into (2.6) gives the more conventional form for orthotropy. The stress-strain relations for isotropic behaviour are obviously given by all the α_i being zero.

The governing equations for anisotropic linear elasticity, just as for any material behaviour, are completed by the equations of motion or, if the inertia terms are negligible, the equilibrium equations

$$\frac{\partial\sigma_{ij}}{\partial x_j} + \rho f_i = 0 \tag{2.7}$$

where ρ denotes the density and $\underset{\sim}{f}$ the body force field.

The literature of the subject includes a number of methods of solution of problems and many particular solutions. Most methods closely follow those already found to be effective in the stress analysis of isotropic materials. Thus for plane deformations,

$$u_1 = u_1(x_1,x_2), \quad u_2 = u_2(x_1,x_2), \quad u_3 = 0,$$

and for transversely isotropic materials with preferred direction parallel to the x_1-axis, the substitution of

$$\sigma_{11} = \frac{\partial^2\chi}{\partial x_2^2}, \quad \sigma_{12} = -\frac{\partial^2\chi}{\partial x_1 \partial x_2}, \quad \sigma_{22} = \frac{\partial^2\chi}{\partial x_1^2}$$

shows that for static problems the Airy stress function χ must satisfy the generalised biharmonic equation

$$A\frac{\partial^4\chi}{\partial x_1^4} + B\frac{\partial^4\chi}{\partial x_1^2\partial x_2^2} + C\frac{\partial^4\chi}{\partial x_2^4} = 0 \tag{2.8}$$

where

$$A = \mu_L c_{11},$$

$$B = (\lambda+2\mu_T)c_{11} - (\lambda+\alpha)^2 - 2\mu_L(\lambda+\alpha), \tag{2.9}$$

$$C = (\lambda+\alpha)\mu_L.$$

By suitably scaling the spatial variables x_1 and x_2 this can obviously be reduced to the usual biharmonic equation of isotropic elasticity, so that again the stresses can be expressed in terms of biharmonic functions. Thus in this case there is very little difficulty added to the analysis by virtue of the material anisotropy. In general, however, the stress analysis of anisotropic materials will be much more complicated than that associated with isotropic behaviour, so that it is more difficult to deduce the implications of an analytical solution when the anisotropy is not weak. Furthermore a study of the literature indicates that the success of some of the methods also depends on the anisotropy being weak, especially those giving series solutions. The term "weakly anisotropic" is here meant to describe behaviour for which the material response does not greatly differ from isotropy, so that a weakly transversely isotropic material has $\mu_L \simeq \mu_T$ and $\alpha, \beta \ll \lambda + 2\mu_T$. Finally, we recall that in isotropic elasticity the usefulness of its solutions often depends on St. Venant's principle whose range of validity extends throughout the body except for a small neighbourhood of the boundary where edge or end effects are important. The principle is often assumed in generalizations to anisotropic elasticity but in fact it can be shown that for strongly anisotropic materials the restriction on its range of validity can render it useless, as we shall discuss later in Sections 6 and 7.

3. FINITE ELASTICITY

The basic theory of finite deformations of an elastic material and a list of appropriate references is treated in detail in one of Professor Rivlin's papers elsewhere in these Proceedings. For anisotropic behaviour the stress-strain relations are

$$\sigma_{ij} = \frac{1}{\sqrt{K_1}} \frac{\partial x_i}{\partial X_R} \frac{\partial x_j}{\partial X_S} \left\{ \frac{\partial W}{\partial G_{RS}} + \frac{\partial W}{\partial G_{SR}} \right\} \qquad (R,S = 1,2,3) \qquad (3.1)$$

where $\underset{\sim}{X}$ denotes the position of a material element in the undeformed configuration, $\underset{\sim}{x}$ its position in the deformed state, W is the strain energy density function, G_{RS} are the components of the Cauchy Green strain tensor

$$G_{RS} = \frac{\partial x_i}{\partial X_R} \frac{\partial x_i}{\partial X_S}$$

and $K_1 = \det \underset{\sim}{G}$.

For <u>transversely isotropic</u> materials with a preferred direction $\underset{\sim}{A}$ in the undeformed configuration, the requirement that W must be form invariant under a rigid rotation of the deformed body implies [4], [6], [7] that W is a function of the five independent invariants

$$\mathrm{tr}\,\underset{\sim}{G}\,, \quad \mathrm{tr}\,\underset{\sim}{G}^2\,, \quad \mathrm{tr}\,\underset{\sim}{G}^3\,, \quad \mathrm{tr}\,(\underset{\sim}{A}\,\underset{\sim}{A}\,\underset{\sim}{G})\,, \quad \mathrm{tr}\,(\underset{\sim}{A}\,\underset{\sim}{A}\,\underset{\sim}{G}^2)\,.$$

An equivalent but more convenient set of strain invariants is

$$\begin{aligned} &J_1 = \mathrm{tr}\,\underset{\sim}{G}\,,\; J_2 = \tfrac{1}{2}\{(\mathrm{tr}\,\underset{\sim}{G})^2 - \mathrm{tr}\,\underset{\sim}{G}^2\},\; J_3 = \mathrm{tr}\,(\underset{\sim}{A}\,\underset{\sim}{A}\,\underset{\sim}{G}^2)\,, \\ &K_1 = \det \underset{\sim}{G}\,,\; K_2 = \mathrm{tr}\,(\underset{\sim}{A}\,\underset{\sim}{A}\,\underset{\sim}{G}) \end{aligned} \qquad (3.2)$$

in terms of which the constitutive equations reduce to

$$\sigma_{ij} = \frac{2}{\sqrt{K_1}}\{(W_1+J_1W_2)g_{ij} - W_2 g_{ik}g_{jk} + W_3(a_i a_k g_{jk} + a_j a_k g_{ik})$$
$$+ K_1 W_4 \delta_{ij} + W_5 a_i a_j\} \qquad (3.3)$$

where

$$W_1 \equiv \partial W/\partial J_1, \quad W_2 \equiv \partial W/\partial J_2, \quad W_3 \equiv \partial W/\partial J_3,$$

$$W_4 \equiv \partial W/\partial K_1, \quad W_5 \equiv \partial W/\partial K_2,$$

$$g_{ij} = \frac{\partial x_i}{\partial X_R}\frac{\partial x_j}{\partial X_R}, \quad a_i = A_R \frac{\partial x_i}{\partial X_R}.$$

Thus here $\underset{\sim}{a}$ represents the current direction and length of a line element which in the undeformed state lies along the preferred direction and is of unit length; g_{ij} are the components of the Finger strain tensor. If $W_3 = W_5 = 0$, (3.3) reduces to the constitutive equation for isotropic materials.

When considering problems involving large deformations of elastic bodies, it is usually assumed that the compressibility is negligible, i.e. the material is effectively incompressible. In this case K_1 is unity, and the stress response must allow an arbitrary hydrostatic pressure $p(\underset{\sim}{x})$ which replaces the term involving W_4.

For an orthotropic material with two preferred directions $\underset{\sim}{A}$ and $\underset{\sim}{B}$ the stress-strain relations may be obtained in a similar, though more complicated, manner and are not quoted here.

A few problems have been solved for these anisotropic materials, usually assuming incompressibility, and most are given in Green and Adkins [7]. They are all examples of controllable deformations [8], which are deformations that are both kinematically admissible and statically admissible (in that they give rise to stresses which satisfy the equilibrium equations). The stress and deformation analyses for boundary value problems involving more general deformations appear to be inordinately difficult.

4. ANISOTROPIC PLASTICITY

The theory of plasticity provides similar difficulties for the solution of boundary value problems involving irreversible plastic behaviour. This is a non-linear theory even for small deformations. Plastic behaviour is characterised by the existence of a yield function (making the crucial distinction between the responses to loading and unloading) and a flow rule which relates the strain increment to the state of stress at a material element. General theories for elastic-plastic continua have been formulated by Hill [9], Pipkin and Rivlin [10] and Green and Naghdi [11]. The theory presented here is different from each of these though in some respects it may well be considered a special case of them.

(i) Yield function

Just as for isotropic metal plasticity, the criterion for plastic yield of an anisotropic material is that a yield function $f(\sigma_{ij})$ exists such that only stress states σ_{ij} for which $f \leqslant 0$ are admissible. If $f = 0$ and $\dot{f} = 0$ the material is in a plastic state; if $f < 0$, or $f = 0$ and $\dot{f} < 0$, it is in an elastic state. Here the superposed dot denotes the operation of time differentiation following a material particle so that if $\underset{\sim}{v}$ represents the velocity at any time t

$$\dot{f}(\underset{\sim}{x},t) = \frac{\partial f}{\partial t} + v_k \frac{\partial f}{\partial x_k} .$$

As is usual in plasticity theory for quasi-static deformations the time t here represents any convenient parameter which determines the sequence of events.

The yield function f in general depends on deformation history as well as on the state of stress, but for convenience we shall not explicitly include this dependence. If the material has any preferred directions then these also must be included in the formulation. We then turn again to the theory of invariants to determine f as a function which is invariant under a rigid body rotation.

Thus for transversely isotropic plastic behaviour with a current preferred direction $\underset{\sim}{a}$ the yield function depends on the stress invariants

$$\mathrm{tr}\,\underset{\sim}{\sigma}\,, \quad \mathrm{tr}\,\underset{\sim}{\sigma}^2\,, \quad \mathrm{tr}\,\underset{\sim}{\sigma}^3\,, \quad \mathrm{tr}\,(\underset{\sim}{a}\,\underset{\sim}{a}\,\underset{\sim}{\sigma})\,, \quad \mathrm{tr}\,(\underset{\sim}{a}\,\underset{\sim}{a}\,\underset{\sim}{\sigma}^2)\,.$$

Furthermore it is usual to assume (again as for isotropic metal plasticity) that yielding is effectively independent of a superposed hydrostatic pressure. Then f depends only on the deviatoric stress

$$\underset{\sim}{s} = \underset{\sim}{\sigma} - \tfrac{1}{3}\underset{\sim}{I}\,\mathrm{tr}\,\underset{\sim}{\sigma} \qquad \text{or} \qquad s_{ij} = \sigma_{ij} - \tfrac{1}{3}\sigma_{kk}\delta_{ij} \tag{4.1}$$

and the invariants are more conveniently written as

$$I_1 = \mathrm{tr}\,\underset{\sim}{s}^2\,,\; I_2 = \mathrm{tr}\,(\underset{\sim}{a}\,\underset{\sim}{a}\,\underset{\sim}{s}^2)\,,\; I_3 = \mathrm{tr}\,\underset{\sim}{s}^3\,,\; I_4 = \mathrm{tr}\,(\underset{\sim}{a}\,\underset{\sim}{a}\,\underset{\sim}{s}) \tag{4.2}$$

with the yield criterion taking the form

$$f(I_1, I_2, I_3, I_4) = 0. \tag{4.3}$$

In the solution of particular problems, even in isotropic plasticity, it is often necessary to adopt a special form for the yield function, for example von Mises's or Tresca's yield function. The von Mises criterion is the most general quadratic isotropic yield function; its generalization to transverse isotropy is therefore

$$f \equiv \frac{1}{2k_T^2} I_1 + \left(\frac{1}{k_L^2} - \frac{1}{k_T^2}\right) I_2 + \frac{1}{Y^2} I_4^2 - 1 = 0 \tag{4.4}$$

where the parameters k_L, k_T and Y are dependent on the deformation history. k_L and k_T may be identified respectively with the yield stresses in shear on planes containing the preferred direction $\underset{\sim}{a}$ in the direction of $\underset{\sim}{a}$ and transverse to $\underset{\sim}{a}$; Y is related to the yield stress σ in uniaxial tension along the preferred direction through the relation

$$\frac{1}{Y^2} = \frac{9}{4\sigma^2} - \frac{1}{k_L^2} + \frac{1}{4k_T^2}\,.$$

These formulations explicitly incorporate the preferred direction $\underset{\sim}{a}$ which can vary from point to point in the body. If $\underset{\sim}{a} = (1,0,0)$, equation (4.4) reduces to the form given by Hill [9]. The von Mises yield criterion for isotropic plastic response is given by (4.4) with $k_L = k_T$ and $I_4 = 0$.

Similar arguments can be used to derive yield criteria for orthotropic materials with two preferred directions $\underset{\sim}{a}$ and $\underset{\sim}{b}$ at any point, and are described in the context of fibre-reinforced materials by Smith and Spencer [12].

(ii) Strain-rates and flow rules

We make the usual assumption of plasticity theory that the strain-rate $\dot{\varepsilon}$ can be expressed as the sum of two parts called the elastic and plastic strain-rates and denoted $\dot{\varepsilon}^e$ and $\dot{\varepsilon}^p$ respectively. Thus

$$\dot{\varepsilon}_{ij} = \frac{1}{2}\left(\frac{\partial v_i}{\partial x_j} + \frac{\partial v_j}{\partial x_i}\right) = \dot{\varepsilon}^e_{ij} + \dot{\varepsilon}^p_{ij} \ . \tag{4.5}$$

For small deformations the elastic strain-rate components $\dot{\varepsilon}^e_{ij}$ are linearly related to the stress-rate components through

$$\dot{\sigma}_{ij} = c_{ijk\ell}\, \dot{\varepsilon}^e_{k\ell} \tag{4.6}$$

with the same elastic constants $c_{ijk\ell}$ as those described previously in Section 2.

For large deformations it is usual to assume that the elastic strain and elastic strain-rates are negligible (i.e. rigid-plastic behaviour). If such an assumption cannot be made, then it is usual to assume the linear relation (4.6). In this case care must be taken to choose a stress-rate measure [13] which is independent of superposed rigid body rotations, such as the Oldroyd derivative of σ_{ij}:

$$\frac{d_O\sigma_{ij}}{d_O t} = \dot{\sigma}_{ij} - \sigma_{ik}\frac{\partial v_j}{\partial x_k} - \sigma_{jk}\frac{\partial v_i}{\partial x_k}$$

which measures the rate of change of the σ_{ij} when these are referred to convected coordinates and measured per current (variable) unit area.

The plastic strain-rate $\dot{\varepsilon}^p$ is given by assuming that the yield function is also a plastic potential function, so that the components are

$$\dot{\varepsilon}^p_{ij} = \begin{cases} 0 & f < 0 \quad \text{or} \quad f = 0,\ \dot{f} < 0 \\ \lambda \dfrac{\partial f}{\partial \sigma_{ij}} & f = 0,\ \dot{f} = 0 \end{cases} \tag{4.7}$$

where λ is a function of deformation history for a work-hardening solid and is an indeterminate factor of proportionality for an effectively non-hardening plastic solid. It is then a straightforward matter to calculate $\dot{\varepsilon}^p_{ij}$ for the yield functions already discussed earlier in this section. It should be noted that the $\dot{\varepsilon}^p_{ij}$ obtained from the deviatoric stress formulation automatically satisfy incompressibility.

Solutions exist for a few problems involving anisotropic rigid-plastic materials. Most are given by Hill [9] who also briefly describes how the conventional slip-line field theory of isotropic materials in plane strain may be generalized to take into account anisotropic mechanical response which can be adequately described by the generalized von Mises yield criterion.

5. STRONGLY ANISOTROPIC MATERIALS AND THE THEORY OF IDEAL COMPOSITES

The preceding sections indicate that in general the stress analysis for an anisotropic material is at least as difficult as that for an isotropic one. However, if the material is strongly anisotropic, then it is in fact possible to construct a simple continuum model of the material which allows us to analyze approximately not only problems which have a complicated "isotropic" analysis but also many which are too difficult to solve if the material were isotropic. We say that the anisotropy is strong if the extensional modulus in some (preferred) direction is much greater than the shear moduli associated with that direction. Thus a strongly transversely isotropic linearly elastic behaviour is characterized by (refer Section 2)

$$\mu_L,\ \mu_T << E.$$

The basis of the simple idealized model was originally introduced [14] in the context of plastic deformations of fibre-reinforced materials. The theory has since been extended to include all types of material response and is most comprehensively described in Spencer's book [3], which also refers to most of the original papers. Another excellent review of the theory has been given by Pipkin [15].

The simplest model - sometimes termed the "ideal fibre-reinforced material" by virtue of the context in which it was proposed and developed - assumes

(i) inextensibility in the preferred ("strong") directions
(ii) that the preferred directions are convected with the material as it deforms
(iii) incompressibility.

The first property is merely the modelling of the kinematical consequence of the material's strength in the preferred direction $\underset{\sim}{a}(\underset{\sim}{X},t)$ - it will shear along or transverse to $\underset{\sim}{a}$ in preference to extension or contraction along $\underset{\sim}{a}$. Thus, making $\underset{\sim}{A}$ a unit vector,

$$|\underset{\sim}{A}(\underset{\sim}{X})| = |\underset{\sim}{a}(\underset{\sim}{X},0)| = |\underset{\sim}{a}(\underset{\sim}{X},t)| = 1. \quad (5.1)$$

From now on $\underset{\sim}{a}$ will be termed the <u>fibre-direction</u>. From the second property we deduce that

$$a_i = A_R \frac{\partial x_i}{\partial X_R} = A_i + A_R \frac{\partial u_i}{\partial X_R} \quad (5.2)$$

so that (5.1) gives

$$A_R A_S \frac{\partial x_i}{\partial X_R} \frac{\partial x_i}{\partial X_S} = 1 \quad (5.3)$$

which for small deformations implies

$$a_i a_j \varepsilon_{ij} = 0. \quad (5.4)$$

The third assumption of incompressibility is the same as that usually made in finite elasticity and plasticity. It should certainly be a reasonable assumption whenever a problem involves large deformations. Even for small deformations there are many quantities of interest whose values are not sensitive to the compressibility of the material. Incompressibility implies

$$\det \underset{\sim}{F} = 1 \quad \text{or} \quad K_1 = 1 \quad (5.5)$$

where $\underset{\sim}{F}$ denotes the deformation gradient tensor with components F_{ij} equal to $\partial x_i/\partial X_j$. For small deformations (5.5) reduces to the condition

$$\varepsilon_{kk} = 0. \quad (5.6)$$

It can be shown [3], [16] that (5.3) and (5.5) imply that the divergences of the initial and current fibre-directions are conserved:

$$\frac{\partial A_R}{\partial X_R} = \frac{\partial a_i}{\partial x_i} \quad (5.7)$$

a result with important implications for the theories of plane or axisymmetric deformations.

For an ideal material with one preferred direction $\underset{\sim}{a}$, the stress tensor can be written as the sum of three distinct parts:

$$\underset{\sim}{\sigma} = T\underset{\sim}{a}\underset{\sim}{a} - p\underset{\sim}{I} + \underset{\sim}{\tau} \tag{5.8}$$

or, in indicial notation,

$$\sigma_{ij} = Ta_i a_j - p\delta_{ij} + \tau_{ij}$$

where T represents the tension in the fibre-direction $\underset{\sim}{a}$ and p the hydrostatic pressure. Both are stress reactions to the kinematical constraints and are independent of the strain field; they are functions of position to be determined by the equilibrium equations.

The extra stress $\underset{\sim}{\tau}$ depends on the state of strain, and constitutive equations are needed to express this dependence. Thus for linear elasticity (2.3), (5.4) and (5.6) show that for transversely isotropic behaviour the stress-strain relations are

$$\sigma_{ij} = Ta_i a_j - p\delta_{ij} + 2\mu_T\varepsilon_{ij} + 2(\mu_L - \mu_T)(a_i a_k \varepsilon_{kj} + a_j a_k \varepsilon_{ki}).$$

It should be noted that now the constitutive equation involves only two elastic constants μ_L and μ_T, which in principle should be able to be determined by two simple shear tests along and transverse to the fibre-direction $\underset{\sim}{a}$. Similarly the ideal model for orthotropic linear elasticity is expressed by

$$\sigma_{ij} = T_a a_i a_j + T_b b_i b_j - p\delta_{ij} + \tau_{ij}$$

and the extra stress τ_{ij} will involve only four independent elastic constants μ, α_1, α_2 and α_3 instead of the thirteen constants for the exact description, and just three ($\alpha_1 = \alpha_2$) instead of nine if $\underset{\sim}{a}$ and $\underset{\sim}{b}$ are mechanically equivalent.

In finite elasticity we can deduce from (3.3) that the ideal model for transverse isotropy is

$$\sigma_{ij} = Ta_i a_j - p\delta_{ij} + 2(W_1 + J_1 W_2)g_{ij} - 2W_2 g_{ik} g_{jk}$$
$$+ 2W_3(a_i a_k g_{jk} + a_j a_k g_{ik}).$$

Even in this case (which is much simpler than the orthotropic case) it is obvious that usually we must make simplifying assumptions about the form of W if we hope to solve any boundary-value problem or even to analyse experimental data obtained on such materials. However, in the particular but very important case of plane deformations in a plane containing the "strong" directions we shall find (Section 6) that such an assumption is not necessary.

The same is true for plastic deformations. In this case the ideal model for a material with one preferred direction [14] is that in general the yield function should depend only on the three stress invariants I_1, I_2 and I_3 (refer (4.2)):

$$f(I_1, I_2, I_3) = 0 \qquad (5.9)$$

and the generalization of von Mises's criterion reduces to

$$f \equiv \frac{1}{k_T^2} I_1 + \left(\frac{1}{k_L^2} - \frac{1}{k_T^2}\right) I_2 - 1 = 0$$

since now $I_4 = 0$. The flow rule will now automatically satisfy the idealization of inextensibility and incompressibility. The yield criterion for a material with two strong directions, with included angle $\psi(\underset{\sim}{x})$ say, has been determined by Smith and Spencer [12]. The simpler quadratic form reduces to

$$f \equiv \frac{I_2+I_4}{C_1^2} + \frac{I_1}{C_2^2} + \frac{I_6}{C_3^2} - 1 = 0$$

where I_1 and I_2 are as before, I_4 and I_6 are given by

$$I_4 = b_i b_j s_{ik} s_{jk}, \qquad I_6 = a_i b_j s_{ik} s_{jk} \cos 2\psi,$$

and C_1, C_2 and C_3 are plastic yield stresses, functions of $\cos^2 2\psi$.

A few non-plane solutions have been obtained (see [3] and [15]) but it is in the theory of plane deformations that the most spectacular success of the model has been demonstrated so far. In the next section we outline that theory, restricting attention to materials with one preferred direction $\underset{\sim}{a}$ or those with two strong directions $\underset{\sim}{a}$, $\underset{\sim}{b}$ which are orthogonal with $\underset{\sim}{b}$ normal to the plane of deformation.

6. PLANE DEFORMATION OF IDEAL ANISOTROPIC MATERIALS

The details of this theory are to be found in [16]. The restriction of plane strain, when added to the other kinematical restrictions of incompressibility and inextensibility, automatically imposes very strong constraints on the possible deformations of the material, and in fact enables one to construct the displacement field for any problem with remarkable ease.

(i) Kinematics

The theory treats problems involving plane deformations of the form

$$x_1 = x_1(X_1, X_2), \quad x_2 = x_2(X_1, X_2), \quad x_3 = X_3$$

and the preferred (inextensible) direction $\underset{\sim}{A}$ in the initial configuration is assumed to lie in the planes X_3 = constant. This unit vector $\underset{\sim}{A}$, if referred to cartesian axes, can be written

$$\underset{\sim}{A} = (\cos\Phi, \ \sin\Phi, \ 0)$$

where Φ is the angle which $\underset{\sim}{A}$ makes with the X_1-axis, and we assume Φ is independent of X_3. Then (5.2) immediately shows that

$$\underset{\sim}{a} = (\cos\phi, \ \sin\phi, \ 0)$$

where ϕ is the angle the inextensible direction makes with the X_1-axis after deformation.

The vector field $\underset{\sim}{a}(\underset{\sim}{X},t)$ defines curves of inextensibility in the material, called <u>fibres</u> or <u>$\underset{\sim}{a}$-lines</u>, which deform as the material deforms, with equations

$$\frac{dx_2}{dx_1} = \tan\phi.$$

The orthogonal trajectories to the fibres are termed <u>normal lines</u> or <u>$\underset{\sim}{n}$-lines</u> and the unit vector field defining the tangents to these lines is given by

$$\underset{\sim}{n} = (-\sin\phi, \ \cos\phi, \ 0).$$

Unlike fibres, normal lines are <u>not</u> material curves in general; a line element lying along a normal line in the undeformed state will ordinarily not lie along a normal line when the body is deformed.

The main kinematical results deduced in [16] are:

(i) For a given particle the curvature κ_n of a normal line passing through it after a deformation is the same as the curvature of the (different) normal line through it before the deformation; i.e.

$$\kappa_n = \kappa_N . \tag{6.1}$$

If the fibres are initially parallel, though not necessarily straight, then

$$\kappa_n = \kappa_N = 0 \tag{6.2}$$

implying that the $\underset{\sim}{a}$-lines will remain parallel throughout any subsequent plane deformation.

(ii) The normal distance between a pair of parallel $\underset{\sim}{a}$-lines is the same all along that pair; furthermore, it is conserved throughout the deformation.

(iii) If the $\underset{\sim}{n}$-curves for parallel fibres intersect along some curve then $\underset{\sim}{a}$ is discontinuous across this curve. In such a case the integral form of (5.7) can be used to show that the curve must bisect the angle between the $\underset{\sim}{a}$-lines on each side of it.

Hence for the case of parallel $\underset{\sim}{a}$-lines these geometrical properties enable a large number of kinematically admissible deformations to be easily constructed. This has considerable practical importance since this case is the one most likely to occur in practice.

Results (i) and (ii) above show that the plane deformation of any element of a body with parallel $\underset{\sim}{a}$-lines is a simple shear along the $\underset{\sim}{a}$-lines, apart from a rigid body rotation (which is taken into account by the directions of $\underset{\sim}{a}$ and $\underset{\sim}{n}$) and a rigid translation which we omit for convenience. The deformation can be represented by

$$\underset{\sim}{F} = (\underset{\sim}{a} + \gamma \underset{\sim}{n})\underset{\sim}{A} + \underset{\sim}{n}\underset{\sim}{N} + \underset{\sim}{k}\underset{\sim}{K} \tag{6.3}$$

where $\underset{\sim}{k}$ and $\underset{\sim}{K}$ $(=\underset{\sim}{k})$ are the unit vector in the X_3-direction and γ is the amount of shear. A further simple result [17] is that if the normal line through a particle makes an angle Θ with some fixed direction before deformation and an angle θ with the same direction after deformation then the amount of shear at that particle is

$$\gamma = \theta - \Theta + c \tag{6.4}$$

where c is constant on each $\underset{\sim}{a}$-line.

(ii) Stress

Since the deformation can be completely specified in terms of γ then the extra stress $\underset{\sim}{\tau}$ can itself only depend on γ or its history:

$$\underset{\sim}{\tau} = \underset{\sim}{\tau}(\gamma). \tag{6.5}$$

With only very weak restrictions on the stress response it can be shown [15], [16] that in fact this extra stress must consist of a shearing stress S (say) and the normal stress required to maintain plane strain conditions. Hence the stress can be written as

$$\underset{\sim}{\sigma} = -p(\underset{\sim}{I} - \underset{\sim}{a}\underset{\sim}{a}) + T\underset{\sim}{a}\underset{\sim}{a} + S(\underset{\sim}{a}\underset{\sim}{n} + \underset{\sim}{n}\underset{\sim}{a}) + S_3\underset{\sim}{k}\underset{\sim}{k} \tag{6.6}$$

where T now represents the <u>total</u> tension on elements normal to the fibre-direction $\underset{\sim}{a}$, p is the total pressure on elements normal to the $\underset{\sim}{n}$-direction and S_3 is a normal stress difference. In the more usual matrix notation the stress can be exhibited as

$$\underset{\sim}{\sigma} = \begin{pmatrix} T & S & 0 \\ S & -p & 0 \\ 0 & 0 & \sigma_{33} \end{pmatrix} \tag{6.7}$$

where the components are referred to <u>local</u> axes coinciding with the directions $\underset{\sim}{a}$, $\underset{\sim}{n}$ and $\underset{\sim}{k}$ respectively. For elastic materials, S and S_3 are functions of γ whose forms can be determined, in principle, from a <u>single simple shearing experiment</u>. For viscoelastic or plastic materials, if the deformation is plane at all times then S and S_3 are determined by the history of γ.

For problems in which the inertia terms may be neglected the stress $\underset{\sim}{\sigma}$ must satisfy the equilibrium equation if it is to be a valid stress solution. Omitting the body force term for convenience, the substitution of (6.6) into (2.7) results in two hyperbolic linear partial differential equations for p and T:

$$\frac{\partial T}{\partial a} = 2\kappa_a S - \frac{\partial S}{\partial n} \tag{6.8}$$

$$\frac{\partial p}{\partial n} - (p+T)\kappa_a = \frac{\partial S}{\partial a}. \tag{6.9}$$

Here $\partial/\partial a$ and $\partial/\partial n$ denote differentiation with respect to distance along the $\underset{\sim}{a}$- and $\underset{\sim}{n}$-lines respectively, and κ_a is the curvature of the $\underset{\sim}{a}$-line at any point.

When a kinematically admissible field has been specified then the fields $\underset{\sim}{a}$ and $\underset{\sim}{n}$ are known, and the shearing stress S can be evaluated from the relevant constitutive equation. Then equations (6.8) and (6.9) can be integrated explicitly along their characteristics, the $\underset{\sim}{a}$- and $\underset{\sim}{n}$-lines; (6.8) gives the tension T and then (6.9) will give p.

If in the deformed body there are regions in which the "fibres" are not only parallel but also straight, then κ_a is zero in these regions. Cartesian coordinates, constant along $\underset{\sim}{a}$-lines and $\underset{\sim}{n}$-lines, can be introduced and both (6.8) and (6.9) integrated directly for T and p. If the $\underset{\sim}{a}$-lines are curved, we can introduce quasi-polar coordinates ξ and θ with ξ as the distance between the $\underset{\sim}{a}$-line and a given reference $\underset{\sim}{a}$-line ($\xi = 0$) and θ the angle that the straight $\underset{\sim}{n}$-line makes with a given direction $\theta = 0$ (as before). Then (6.8) and (6.9) can again be explicitly integrated to give

$$T = \int_0^{\theta} \left(r_a \frac{\partial S}{\partial \xi} + 2S \right) d\theta' + T_0(\xi)$$

$$r_a p = r_0(\theta) p_0(\theta) - \int_0^{\xi} \left(T + \frac{\partial S}{\partial \theta} \right) d\xi' \qquad (6.10)$$

where r_a denotes the radius of curvature at a point and $r_0(\theta)$ the radius of curvature of the reference $\underset{\sim}{a}$-line $\xi = 0$:

$$r_a = 1/\kappa_a = r_0(\theta) + \xi.$$

We note that (6.10) shows that the value of T may be specified at one and only one point on each $\underset{\sim}{a}$-line, and that of p at only one point on each $\underset{\sim}{n}$-line. This means that if a particular $\underset{\sim}{a}$-line crosses the boundary of a body twice, then the component of stress in the $\underset{\sim}{a}$-direction cannot be specified arbitrarily at both boundary points; a similar restriction applies to boundary values of p. This is in sharp contrast with the more familiar theories of elasticity, in which the stress can be specified arbitrarily at all points of the boundary of an elastic body. Similarly, the displacement field cannot usually be arbitrarily specified at two points of the same $\underset{\sim}{a}$-line or $\underset{\sim}{n}$-line in an ideal material. Mathematically, this contrast is a consequence of the properties of the elliptic partial differential equations of elasticity as against the hyperbolic equations of the ideal theory. Physically it implies that for strongly anisotropic materials the conditions at one point of a body strongly affect the conditions at large distances along an $\underset{\sim}{a}$-line or $\underset{\sim}{n}$-line passing through that point. Thus end effects are not now localized to one or two typical diameters' distance from the boundary, but can possibly extend throughout the body. Hence St. Venant's principle is now violated in many problems for which it could be assumed in isotropic elasticity.

Another remarkable result of the theory when compared with the theory of isotropic materials is that it predicts the possibility of thin layers of very high stress occurring along and normal to the $\underset{\sim}{a}$-direction. These layers (which are lines in the plane of deformation but extend in the X_3-direction) occur not only when concentrated loads are applied to the boundary but also whenever a free flexed surface of a body contains the "fibres" or normals and does not intersect them. They can also occur inside the body. This contrasts with the conventional isotropic theories which usually allow only point or line singularities in stress, such as at the tip of a crack in plane strain or plane stress.

These properties are a feature of the "ideal" theory and indicate that the details of boundary conditions are much more important in strongly anisotropic materials than they are for isotropic bodies.

Many problems of a practical nature have now been solved using the theory. These include the bending of laminated [18] and fibre-reinforced beams [19], the compression of a hollow reinforced tube (the radial tyre problem) [20] and the solution of a number of crack problems [21], [22].

The equivalent theory for materials which are strong in two directions has also been developed, and most of the work is included in Spencer's book [3], and will not be treated here.

7. EXTENSION OF THE THEORY TO REAL MATERIALS

Of course, no real material is either exactly incompressible or inextensible in any direction, let alone both. Also, since the ideal theory sometimes predicts these layers of extremely large stress it might be thought that the theory must be suspect, and at best irrelevant to real problems involving real materials. However, there exist a number of exact solutions in the theory of linear anisotropic elasticity, and these solutions can be investigated to see in each case whether the equivalent solution given by the ideal theory bears any relevance to the exact solution.

This was first done by Everstine and Pipkin [23] who considered some simple but illuminating examples of plane deformations of a strongly anisotropic (but not inextensible or incompressible) material. This involved solving equation (2.8) for the Airy stress function χ subject to various boundary conditions which were deliberately chosen so as to enable an exact solution to be found. They showed that the singular layers of the ideal theory become regions of stress concentration which attenuate slowly with distance along or normal to the preferred direction. The solutions also provided estimates for the stress in and dimensions of these layers.

A typical example is that of the half-space $x_2 \geqslant 0$ with boundary conditions

$$\sigma_{12} = \sigma_0 \cos(x_1/\ell), \quad \sigma_{22} = 0 \qquad \text{on } x_2 = 0$$

$$\sigma_{ij} \to 0 \qquad \text{as } x_2 \to \infty. \tag{7.1}$$

The preferred strong direction at each point is parallel to the x_1-axis.

The idealized theory gives a trivial solution for the displacements:

$$u_1 = 0, \quad u_2 = 0, \quad u_3 = 0$$

so that σ_{12} (=S) is zero inside the material, but non-zero on the boundary. The equilibrium equations then give σ_{22} (= -p) to be zero while σ_{11} (= T) is given in terms of the derivative of the discontinuous shear stress:

$$\sigma_{12} = 0 \quad \text{for } x_2 > 0, \qquad \sigma_{12} = \sigma_0 \cos(x_1/\ell) \quad \text{at } x_2 = 0$$

$$\sigma_{22} = 0, \qquad \sigma_{11} = \sigma_0 \ell \sin(x_2/\ell)\,\delta(x_2) \tag{7.2}$$

where $\delta(x_2)$ denotes the Dirac delta function.

The exact anisotropic solution is easily obtained and is given in [23]. For a strongly anisotropic material an adequate but more illustrative approximation to that solution is

$$\sigma_{12} = \sigma_0 \cos(x_1/\ell) e^{-x_2/\ell\varepsilon_t} + O(\varepsilon_c \varepsilon_t)$$

$$\sigma_{22} = O(\varepsilon_t) \tag{7.3}$$

$$\sigma_{11} = \sigma_0 \ell \sin\left(\frac{x_1}{\ell}\right) \frac{1}{\ell\varepsilon_t} e^{-x_2/\ell\varepsilon_t} + O(\varepsilon_t)$$

where ε_t and ε_c are parameters related to the elastic constants and are measures of extensibility and compressibility respectively:

$$\varepsilon_t^2 \simeq \mu_L/E \ll 1, \qquad \varepsilon_c^2 \simeq 2\mu_L(1-\nu')/E' \ll 1.$$

A study of (7.3) immediately shows that as ε_t and ε_c tend to zero, the stress distribution tends to that given by the "ideal" theory in (7.2). The singular stress layer σ_{11} is seen to be a boundary layer region of thickness of the order of ε_t times the characteristic length ℓ, in which the stress in the preferred direction is very large, of the order of $\sigma_0/\ell\varepsilon_t$.

Similar results were obtained for normal lines. It was also shown that stresses imposed in these characteristic directions decayed slowly, penetrating to a distance of order ℓ/ε; this result supports the contention made in the previous section concerning St. Venant's principle.

Further development of this work by the same authors [24] and others [25], [26] has now fully substantiated the value of the idealized theory. It also shows, at least for linear elasticity, how the ideal solution may be used in conjunction with the boundary layer analysis [24], [26] to give meaningful analytical results for strongly anisotropic materials in boundary value problems for which no analytical results are even known for isotropic materials.

Although nearly all the work with strongly anisotropic materials has so far concerned static or quasi-static problems, a start has been made in extending the theory to dynamic plasticity [27]. It is easy to predict that the ideal theory will again highlight the unusual behaviour to be expected due to the strong anisotropy, with the deformation again dominated by shearing rather than extensional modes of propagation.

Finally, it should be reiterated that whilst the idealized theory and its extensions are appropriate for strongly anisotropic behaviour, they should not be expected to provide any information about weakly anisotropic bodies, the theory of which is still a subject of considerable research.

REFERENCES

1. I. S. Sokolnikoff, Mathematical Theory of Elasticity, 2nd edition, McGraw-Hill, New York-Toronto-London, 1956.
2. A. E. H. Love, The Mathematical Theory of Elasticity, 4th edition, Cambridge University Press, 1952.
3. A. J. M. Spencer, Deformations of Fibre-reinforced Materials, Oxford Science Research Papers, Clarendon Press, Oxford, 1972.
4. A. J. M. Spencer, Theory of invariants, in: Continuum Physics, Vol.1, ed. by A. C. Eringen, Academic Press, New York and London, 1971.
5. S. G. Lekhnitskii, Theory of Elasticity of an Anisotropic Body, Holden-Day, San Francisco, 1963.
6. J. L. Ericksen and R. S. Rivlin, J. rat. Mech. Analysis 3, 281 (1954).
7. A. E. Green and J. E. Adkins, Large Elastic Deformations, Clarendon Press, Oxford, 1960.
8. A. C. Pipkin, Non-linear phenomena in continua, in: Non-Linear Continuum Theories in Mechanics and Physics and their Applications, Edizioni Cremonese, Rome, 1970.
9. R. Hill, The Mathematical Theory of Plasticity, Clarendon Press, Oxford, 1950.
10. A. C. Pipkin and R. S. Rivlin, Z. angew. Math. Phys. 16, 313 (1965).
11. A. E. Green and P. M. Naghdi, Arch. ration. Mech. Analysis 18, 251 (1965).
12. G. E. Smith and A. J. M. Spencer, Q. Jl Mech. appl. Math. 23, 489 (1970).
13. J. F. Mulhern, T. G. Rogers and A. J. M. Spencer, Int. J. Eng. Sci. 7, 129 (1969).
14. J. F. Mulhern, T. G. Rogers and A. J. M. Spencer, Proc. R. Soc. A 301, 437 (1967).
15. A. C. Pipkin, Finite deformations of ideal fiber-reinforced composites, in: Micromechanics, ed. by G. P. Sendeckyi, Academic Press, New York and London, 1974.
16. A. C. Pipkin and T. G. Rogers, J. appl. Mech. 38, 634 (1971).
17. A. C. Pipkin and T. G. Rogers, Q. appl. Math. 29, 151 (1971).
18. A. J. M. Spencer, Q. Jl Mech. appl. Math. 25, 387 (1972).
19. T. G. Rogers and A. C. Pipkin, J. appl. Mech. 38, 1047 (1971).
20. T. G. Rogers and A. C. Pipkin, Q. Jl Mech. appl. Math. 24, 311 (1971).
21. A. H. England and T. G. Rogers, Q. Jl Mech. appl. Math. 26, 303 (1973).

22. J. N. Thomas and A. H. England, J. Inst. Maths Applics 14, (in press 1974).
23. G. C. Everstine and A. C. Pipkin, Z. angew. Math. Phys. 22, 825 (1971).
24. G. C. Everstine and A. C. Pipkin, J. appl. Mech. 40, 518 (1973).
25. A. H. England, J. E. Ferrier and J. N. Thomas, J. Mech. Phys. Solids 21, 279 (1973).
26. A. J. M. Spencer, Int. J. Solids and Structures, (in press 1974).
27. A. J. M. Spencer, J. Mech. Phys. Solids 22, 147 (1974).

SYMMETRIC MICROMORPHIC CONTINUUM: WAVE PROPAGATION, POINT SOURCE SOLUTIONS AND SOME APPLICATIONS TO EARTHQUAKE PROCESSES

Roman Teisseyre

Institute of Geophysics,
Polish Academy of Sciences
02-093 Warszawa, Pasteura 3, Poland

1. INTRODUCTION

Our considerations are pointed to a role of microstructure in a geological space and especially in that region of Earth interior where earthquakes are occuring. It seems that internal structure and discontinuities have an essential influence on development of geophysical processes. Moreover, it is resonable to believe that the conception of continuum with microstructure can be a suitable tool to describe the earthquake phenomena and wave propagation in the near focal zone.

The scale of Earth structures with their extremely complicated internal microstructures justifies the use of continuum mechanics and especially that of generalized micromorphic continuum.

We will follow in general the approach to micromorphic continuum as developed in the papers of Eringen /1968/; Eringen, Claus /1970/; Eringen, Suhubi /1964/; Suhubi, Eringen /1964/.

It is a general point of view that earthquake processes depend both on the state of stresses and the structure of the focal zone. However, this postulate has not been used in the explicite way. The point force model of an earthquake source is localized somewhere in the space, assuming silenciously that a certain inhomogeneity creates the attaching point to release the internal stresses. Better understanding brought the dislocation and crack models. These material defects constitute the objects of dynamical interractions and enable to define the earthquake process and its

Thoft-Christensen (ed.), Continuum Mechanics Aspects of Geodynamics and Rock Fracture Mechanics, 201-244.

development. A complicated structure of the focal zone establishes the appropriate conditions in which stresses form the dislocation field and in which the earthquake processes can develop and can be described by a fairly adequate theory. It is believed that such theory could be a micromorphic one.

2. BASIC ASSUMPTIONS

Earthquake source models developed by different authors can be divided into three groups. First group deals with body force distribution in a focus region, second presents dislocation processes as responsable for the internal energy releases in earthquakes, and in the third group the crack mechanics is considered in order to describe fracturing in an earthquake zone.

First and second groups are mutually connected by the theorem of stress equivalency caused by dislocation and by point force distribution. Second and third groups join the theory of continuous distribution of dislocations. Special cases of such distributions correspond to fields of cracks.

Dislocational approach introduces some concepts of medium inhomogeneities while cracks described further stage that means rupturing and lost of cohesion in a part of medium.

The new model based on theory of micromorphic continuum introduces microstructure and its deformations into the earthquake processes occuring in the focal regions. This model is called here the earthquake structure model.

The basic assumption of the earthquake structure model concerns the existence of the small structural elements permeating focal region. We can imagine that some processes in the past have changed the structure of the medium; either some sort of metamorphic processes or phase transformations or some more mechanical processes have caused the non-reversible changes along certain elements which could suffered fracturings and displacements. We can think also on grains in the mineralogical sense or - in a greater scale - on a block structure constituting fractured tectonical zone. These internal microstructural elements form the attaching points around which the stresses concentrate.

Further on, in applications, the simple model of earthquake structure will be discussed; it will be assumed that the above mentioned structural elements permeating focal region may single a certain direction out. This kind of microanisotropy will be expressed in the calculations by means of the anisotropy of micro-inertia tensor only.

To complete our model of earthquake structure we demand that deformations imposed on microstructural elements favour their elongation rather than rotation. This justifies our choice of micromorphic continuum in which length deformations of Cosserat directors are allowed. The above formulated condition limiting rotations of microelements will be ensured by a suitable choice of material constants in the constitutive equations.

The theory of continuum with microstructure presents a great interest for seismology. It seems that we can thus approach closer to the real conditions of structures and their deformations in the Earth. The deformations in a micromorphic theory are represented not only by the displacement vector $\underset{\sim}{u}$, but here enters a new tensor which describes deformations and rotations of microelements /grains, blocks or some internal surface defects/. This is a microdisplacement tensor $\underset{\sim}{\phi}$.

The deformation can be now expressed by:

strain tensor: $E_{ik} = \frac{1}{2}/u_{i,k} + u_{k,i}/$

microstrain tensor: $\varepsilon_{nl} = u_{l,n} + \phi_{nl}$

microdislocation density: $A_{kl} = -e_{lmn}\,\phi_{kn,m}.$

Vector and tensor notation is used here parallely.

The stress measures are described by the stress tensor $\underset{\approx}{T}$ /not necessarily symmetric/, the microstress tensor $\underset{\approx}{S}$ and the stress moment tensor $\underset{\sim}{\Lambda}$.

The relation between strain measures and stresses defines material properties. In the micromorphic theory the constitutive equations contain a greater number of material constants. Together with the Lamé λ, μ we will have the following ten constants: τ, σ, η, k, ν, a, b, c /Teisseyre, 1972/.

The theory is considerably simplified assuming rotations of grains only. In this case the microdisplacement tensor can be reduced to a vector of rotations /micropolar or Cosserat medium/.

For seismological problems, especially when considering a focal region and its deformations, a micropolar theory is rather not adequate.

The model of an earthquake zone is thus represented here by micromorphic continuum assuming special conditions put on the couples. It shall be mentioned here that the external action can be here represented not only by the body forces $\underset{\sim}{f}$ but also the body couples $\underset{\sim}{L}$.

The assumptions are now the following:

a/ The rotation moment formed by the divergence of the stress moment vanishes:

$$\Lambda_{plk,p} = \Lambda_{pkl,p} \text{ that is } \Lambda_{p[lk],p} = 0.$$

This condition means that the stress moment tensor forms a zero rotation moment on a volume element. This condition can be satisfied identically by a suitable choice of the material constants in the constitutive equations. In this case ten unknow constants are reduced to the quantity of seven. These constants shall be chosen in such a way as to describe a complicated internal structure of a focal region.

b/ The theories of micropolar or micromorphic continuum are called sometimes theories of nonsymmetric stresses. This is so because in these theories the symmetry condition for stress tensor $\underset{\sim}{T}$ does not hold. However, here we put the next condition that the stress tensor is symmetric:

$$T_{ik} = T_{ki}.$$

c/ The rotation moment of body couples vanishes:

$$L_{ik} = L_{ki}, \text{ that is } L_{[ik]} = \frac{1}{2}\left[L_{ik} - L_{ki}\right] = 0.$$

This condition expresses that the external action can be exerted by stresses surrounding the focal region.

d/ In the balance equations of the micromorphic theory there enter also some other quantities; we shall mention here a tensor of microinertia $\underset{\sim}{I}$. The microelements differ in general by their density ϱ' in comparison with the base density ϱ. We assume that only diagonal elements of a microinertia tensor differ from zero:

$$I_{ik} = \delta_{ik} I_{ii} \text{ /not summed/}.$$

We shall remind here the definition of microinertia tensor. Let us take the component I_{xx}:

$$I_{xx} = \frac{\varrho'}{\varrho} \lim_{v \to 0} \frac{\sum_i \Delta v^{(i)}}{\Delta V} \left/ \bar{y}^2_{(i)} + \bar{z}^2_{(i)} \right/$$

where: $\Delta v^{(i)}$ - the microelement volumes in an element ΔV,

$/\bar{y}_i^2 + \bar{z}_i^2/^{1/2}$ - the distances to a mass center of ΔV.

The microinertia tensor plays also an essential role in the deformation pattern. It shall be noted here that the well-known rigid body inertia tensor J_{kl} is formaly related to I_{kl} by the relation:

$$J_{kl} = I_{nn}\,\delta_{kl} - I_{kl}$$

The above formulated assumptions a - d define the symmetric micromorphic continuum.

The first two assumptions are of the axiomatic character, while next ones concern the physical circumstances of the processes involved.

3. A GEOMETRICAL MEANING OF STRAIN MEASURES AND THE COMPATIBILITY EQUATIONS

It has been said that in the generalized continuum theories the strains are given not only by the displacement vector $\underset{\sim}{u}$ but also by the microdisplacement tensor $\underset{\sim}{\phi}$. In the case of micropolar medium

$$\phi_{ns} = -\,\phi_{sn};$$

three independet components define microrotations of Cosserat directors.

It has been said before that material description will be reffered here to a certain reference configuration of the body in which the indexes of components are marked by majuscule latin letters. In a spatial description we use for indexes the small latin letters. The deformation of the body is thus described at time t by the mapping:

$$x_n = x_n/x_K,\ t/ \qquad /1/$$

We assume that the inverse motion is uniquely determined.

In the micromorphic medium it is assumed that the body posses certain granular structure. We assume that a volume element $/\Delta V + \Delta S/$ of the body contains N discrete grains /microvolumes $/\Delta v^{(i)} + \Delta s^{(i)}$ // with a mass density ϱ'/. Prior to deformation a material point is described by

$$x_K^{o(i)} = x_K^o + \xi_K^{o(i)} \qquad /2/$$

where: $\xi^{o(i)}$ defines a position of a material point of the microvolume in relation to a center of mass of the volume x_K^o.

After deformation the corresponding spatial description reads

$$x_n^{(i)} = x_n + \xi_n^{(i)}$$

The axiom of affine motion defines the relative deformation functions K_{nN} /x,t/:

$$\xi_n^{(i)} = K_{nN}/x,t/\ \xi_N^{o(i)} \qquad /3/$$

The inverse micromotion K_{Nn} satisfies the relations

$$K_{nN} K_{Ns} = \delta_{ns}\ , \quad K_{nK} K_{Ln} = \delta_{KL}$$

In the classical theory the deformation is given by displacement vector

$$u_n = x_n - x_N^o \delta_{Nn}\ , \quad u_N = x_n \delta_{Nn} - x_N^o$$

hence the deformation gradients:

$$x_{n,N} = /\delta_{LN} + u_{L,N}/\delta_{nL}\ , \quad x_{N,n} = /\delta_{ln} - u_{l,n}/\ \delta_{Nl}$$

In analogy to the above we introduce after Suhubi, Eringen /1964/ the microdisplacement tensor ϕ_{KL} - material description, ϕ_{kl} - spatial description:

$$K_{nN} = /\delta_{LN} + \phi_{LN}/\delta_{nL}\ , \quad K_{Nn} = /\delta_{ln} - \phi_{ln}/\delta_{Nl} \qquad /4/$$

Further on we will limit our considerations to a linear theory. We have:

$$\phi_{KL} = \phi_{kl}\ \delta_{kK}\ \delta_{lL}$$

The following strain measures are now defined /material and spatial descriptions/:

$$\left.\begin{array}{ll}\left.\begin{array}{l}E_{NL} = u_{(N,L)} \;, \\ E_{nl} = u_{(n,l)}\end{array}\right\} & \text{/strain tensor/} \\ \left.\begin{array}{l}\mathcal{E}_{NL} = u_{L,N} + \phi_{NL} \;, \\ \mathcal{E}_{n,l} = u_{l,n} + \phi_{nl}\end{array}\right\} & \text{/microstrain tensor/} \\ \left.\begin{array}{l}\Gamma_{KLM} = -\phi_{KL,M} \;, \\ \Gamma_{klm} = -\phi_{kl,m}\end{array}\right\} & \text{/microstrain tensor/}\end{array}\right\} \quad /5/$$

The following compatibility conditions follow from the above definitions:

$$\left.\begin{array}{l}\epsilon_{kmp}\,\epsilon_{lnq}\,E_{mn,pq} = 0 \\ \epsilon_{kmn}\,/\mathcal{E}_{ml,n} + \Gamma_{mln}/ = 0 \\ \epsilon_{kpq}\,\Gamma_{lmp,q} = 0\end{array}\right\} \quad /6/$$

With definitions /5/ and relation /4/ it is easy to express the difference between squares of arc length after and before deformation /linear theory/:

$$\begin{aligned} ds^2 - dS^2 \approx\ & 2/E_{KL} + \Gamma_{KML}\,\xi^o_M/dx^o_K\,dx^o_L \\ & + 2/\mathcal{E}_{KL} + \Gamma_{LMK}\,\xi^o_M/dx^o_K\,d\xi^o_L \\ & + 2/\mathcal{E}_{(KL)} - E_{KL}/d\xi^o_K\,d\xi^o_L \end{aligned}$$

For the linear theory the volume changes are given as follows:
macrovolume:

$$\frac{dV}{V_0} \approx E_{KK} + \Gamma_{KLK}\,\xi^o_L$$

microvolume:

$$\frac{dv}{v_o} \approx \phi_{KK}$$

These expressions can be directly rewritten in the spatial description. At the end of this résumé of micromorphic mechanics we will give the expression for relative velocities of a material point $x_k + \xi_k$ with respect to the center of mass x_K.

According to /3/ we get:

$$\dot{\xi}_n = \dot{K}_{nN} \xi^o_N$$

and expressing ξ^o_N with help of the inverse motion K_{Nn}

$$\dot{\xi}_n = V_{ns} \xi_s$$

where:

$$V_{ns} = \dot{K}_{nN} K_{Ns}$$

V_{ns} is called the gyration tensor; now we can get easily the important relation valid in the linear approximation:

$$V_{ns} = \dot{\phi}_{ns} \qquad /7/$$

4. MOTION EQUATIONS

The balance laws given by Eringen, Claus /1970/ for a micromorphic continuum are as follows:

$$\dot{\varrho} + /\varrho v_k/,_k = 0 \quad \text{/mass conservation/} \qquad /8/$$

$$\dot{I}_{kl} + I_{kl,m} v_m - I_{lm} V_{km} - I_{km} V_{lm} = 0 \qquad /9/$$

/microinertia balance/

$$T_{kl,k} + \varrho f_l = \varrho \dot{v}_l \quad \text{/balance of momentum/} \qquad /10/$$

$$\Lambda_{plk,p} + /T_{kl} - S_{kl}/ + \varrho L_{lk} = \varrho \dot{\Sigma}_{lk} \qquad /11/$$

/balance of moment of momentum/

$$T_{kl} v_{l,k} - /T_{kl} - S_{kl}/V_{lk} + \Lambda_{klm} V_{lm,k} \qquad /12/$$

$$+ q_{k,k} + \varrho h = \varrho \dot{\epsilon} \quad \text{/energy conservation/}$$

We did not mentioned yet some quantities appearing in the eqs. /8-12/: $\underset{\sim}{v}$ - velocity, $\underset{\sim}{\Sigma}$ - inertia spin tensor, q - heat flux vector, ϵ - density of internal energy, h - density of heat production.

In the further calculations we will confine ourselves only the linear theory and Cartesian coordinate system.

The inertia spin tensor $\underset{\sim}{\Sigma}$ can be expressed by means of microinertia tensor $\underset{\sim}{I}$ and gyration tensor $\underset{\sim}{V}$

/Eringen, Claus 1970/:

$$\dot{\Sigma}_{lm} = I_{sm} /\dot{v}_{ls} + v_{lp} v_{ps}/$$

In the linear theory:

$$\dot{\Sigma}_{lm} = I_{sm} \ddot{\phi}_{ls} \qquad /13/$$

The equations of motion shall be completed by the constitutive equations which join the strain measures with stress tensors for a given type of material constituting the continuum considered.

5. DISLOCATIONS IN THE MICROMORPHIC CONTINUUM

Following Eringen, Claus /1970/ we will discuss the definition of dislocation density in the micromorphic continuum.

Kröner's /Kröner, Seeger, 1959/ approach to dislocations is based on the incompatibility relation. Total strains can be split into elastic and plastic parts. At the same time total strains shall satisfy the compatibility conditions. These conditions thus impose the relations for elastic and plastic parts of strains; the plastic part grouped on one side of the equations define the dislocation density tensor. A similar procedure could be applied to microstrain measures. The compatibility conditions /6/ define relations in which the plastic part of microstrains define micromorphic dislocation density and disclination density /density reffered to rotational dislocations/.

$$\Pi_{kl} = -\epsilon_{kmn} \phi^{plastic}_{ml,n} + \epsilon_{kmn} \Gamma^{plastic}_{mln} \qquad /dislocations/$$

$$\Theta_{kml} = -\epsilon_{kns} \Gamma^{plastic}_{mln,s} \qquad /disclinations/$$

These definitions agree with those for micropolar media given by Eringen, Claus /1970/.

Regarding Γ not as the independent microstrains but related to $\tilde{\phi}$ according to /5/ we get

$$\frac{1}{2}\Pi_{kl} = -\epsilon_{kmn} \phi^{plastic}_{ml,n} , \quad \Theta_{kml} = 0$$

The last formula with $\frac{1}{2}\tilde{\Pi}$ corresponds to Kröner's definition of dislocation density $\tilde{A}$ if the plastic distortions $\tilde{U}$ /second order tensor/ are introduce besides the elastic displacements $\tilde{u}$ /Teisseyre, 1969/.

The microstructure of medium includes, however,

directly a possibility of appearance of dislocations even if we exclude the plastic phenomena.

In this paper the earthquake structure model is related to microdislocation distribution, bounded to fine structure of focal region. This fine structure can be formed by some elements of the fracture planes or by certain distribution of blocks or grains. Moreover, we will not split, in a heuristic way, the deformations into the elastic and plastic parts, as it seems that the plastic phenomena are to some extent included here through micromorphic or, one can said, granular structure of the medium.

Indeed, we can define an average Burgers vector taking integral over a closed circuit bounding a small surface element /Eringen, Claus, 1970/:

$$\oint d\xi_n = \oint K_{nN}\, d\xi^o_N = b_n \Delta s$$

The corresponding definition in material frame can be set up. For a dislocation density /spatial description/ follows - after transformation of this integral to a surface one - the relation

$$A_{kl} = K_{kK}\, \epsilon_{lmn}\, K_{Kn,m}$$

or in linear terms according to /4/ and /5/:

$$A_{kl} = -\epsilon_{lmn} \phi_{kn,m} = \epsilon_{lmn} \Gamma_{knm} \qquad /14/$$

This definition of $\underset{\sim}{A}$ can be also immediately obtained from the formula given by Kröner and Seeger /1959/:

$$A_{kl} = -\epsilon_{lmn} \bar{\Gamma}_{[mn]k}$$

with an appropriately defined affine connection $\bar{\Gamma}_{nmk}$ /Eringen, Claus, 1970/:

$$\bar{\Gamma}_{nmk} = K_{kK}\, K_{Kn,m}$$

The antisymmetric part of $\bar{\Gamma}_{nmk}$ is known as torsion tensor:

$$\bar{\Gamma}^k_{[ml]} = S^k_{ml}$$

Returning now to the definitions /5/ we come back again to /14/ in linear approximation. However, in general case, in the relations /4/ the Krönnecker deltas shall be replaced by the appropriate metric tensors.

The Riemann curvature tensor $\underset{\sim}{R}$ can be expressed by Christoffel symbols and torsion $\underset{\sim}{S}$

$$R_{nmlk} = 2\,\bar{\bar{\Gamma}}_{[m|lk|,n]} - 2\,g^{pq}\,\bar{\bar{\Gamma}}_{[m|kq|}\bar{\bar{\Gamma}}_{n]lp} = 0$$

where:

$$\bar{\Gamma}^{k}_{ml} = \left\{ {k \atop ml} \right\} + S_{lm}^{\cdot\cdot k} - S_{m\cdot l}^{\cdot k} + S^{k}_{\cdot lm}$$

The curvature tensor, of course, vanishes, also vanishes the corresponding Einstein tensor

$$N^{ts} = \frac{1}{4}\epsilon^{tlm}\,\epsilon^{spq}\,R_{pqlm} = 0$$

We can, however, split $\underset{\sim}{R}$ or better $\underset{\sim}{N}$, in two parts, such that one of them is formed by metric tensor and the other contains torsion and - in nonlinear theory - a metric tensor too. This second part is related to the incompatibility properties of the continuum. In the linear approximation we get /Kröner, Seeger, 1959/:

$$\epsilon_{jnm}\,\epsilon_{ilk}\,E_{mk,nl} - \epsilon_{jnm}\,A_{mi,n} = 0$$

The density of disclinations is defined by the second term and it corresponds to density of rotational dislocations defined by Teisseyre /1969/:

$$\tilde{N}_{ji} = \epsilon_{jnm}\,A_{mi,n}$$

The symmetric part of $\tilde{N}$ is the Kröner incompatibility tensor.

Taking into account the relation for density of microdislocations we get

$$\tilde{N}_{ji} = \epsilon_{jsk}\,\epsilon_{imn}\,\phi_{kn,ms}$$

or assuming $\underset{\sim}{\Gamma}$ as an independent strain measure:

$$\tilde{N}_{ji} = -\epsilon_{jsk}\,\epsilon_{imn}\,\Gamma_{knm,s}$$

A systematic derivation of nonlinear theory of generalized medium has been presented by Ben-Abraham /1970/. The total deformation is divided in three parts elastic, plastic /dislocations/ and qlastic /extramatter/. The corresponding distortions - joining the linear elements of two nearby material points referred to the different states - enable to define the appropriate strain measures. The linear connexions constructed with help of those distortions can be defined by a formula of the type:

$$\bar{\Gamma}^{\mu}_{\alpha\beta} = K^{\mu}_{K} K^{K}_{\beta,\alpha}$$

The fundamental geometric equation states that the curvature tensor associated with total deformation vanishes. The non-Riemannian part of the curvature tensor consists two tensors which describe the distribution of defects. The torsion tensor describes the dislocation distribution and the ammetry tensor the extramatter distribution. Ben-Abraham proved that the micromorphic theory follows from this general consideration.

6. THE CONSTITUTIVE EQUATIONS AND THE ENERGY DENSITY FUNCTION

In this part of this paper we will proceed with the constitutive equations which define the particular type of micromorphic medium that is which define the relation between stresses $\underset{\sim}{T}$, $\underset{\sim}{S}$, $\underset{\sim}{\Lambda}$ and strain measures $\underset{\sim}{E}$, $\underset{\sim}{\mathcal{E}}$ and $\underset{\sim}{\Gamma}$ or $\underset{\sim}{E}$, $\underset{\sim}{\mathcal{E}}$ and dislocation density tensor $\underset{\sim}{A}$.

The nonlinear constitutive equations for microelastic, isotropic solids have been presented by Suhubi and Eringen /1964/. For the linear theory they are reduced to:

$$\begin{aligned} T_{kl} &= /\lambda + \tau/E_{nn}\delta_{kl} + 2/\mu + \sigma/E_{kl} + \\ &+ \eta\,\mathcal{E}_{nn}\delta_{kl} + k\mathcal{E}_{kl} + \nu\,\mathcal{E}_{lk} \\ S_{kl} &= /\lambda + 2\tau/E_{nn}\delta_{kl} + 2/\mu + 2\sigma/E_{kl} + \\ &+ /2\eta - \tau/\mathcal{E}_{nn}\delta_{kl} + /k + \nu - \sigma/\ /\mathcal{E}_{kl} + \mathcal{E}_{lk}/ \\ \Lambda_{plk} &= aA_{rn}\ /\epsilon_{prn}\delta_{kl} - \epsilon_{krn}\delta_{pl}/ - b\epsilon_{pkn}A_{ln} + \\ &+ c/\epsilon_{pln}A_{kn} - \epsilon_{kln}A_{pn}/ \end{aligned} \qquad /15/$$

According to the condition b/ - symmetry of the stress tensor $\underset{\sim}{T}$ - we shall put:

$$\nu = k \qquad /16/$$

The condition a/ states that the difference between the divergences of the stress moment components vanishes. This implies for the constants a and b :

$$a = 2c, \quad b = -c \tag{17}$$

It shall be reminded that the earthquake structure model includes a possibility of microanisotropy of the focal region. A microanisotropy can be introduced through microinertia tensor $\underset{\sim}{I}$. To get the energy density function we will first rewrite the motion equation in the vector form:

$$\operatorname{div} \underset{\sim}{T} - \varrho \ddot{\underset{\sim}{u}} = -\varrho \underset{\sim}{f}$$

$$\operatorname{div} \underset{\sim}{\Lambda} + (\underset{\sim}{T} - \underset{\sim}{S}) - \varrho \ddot{\underset{\sim}{\phi}} \underset{\sim}{I} = -\varrho \underset{\sim}{L}$$

Starting with these equations and the virtual work principle

$$\int \left[(\varrho \underset{\sim}{f} - \varrho \ddot{u}) \delta \underset{\sim}{u} + (\varrho \underset{\sim}{L} - \varrho \ddot{\underset{\sim}{\phi}} \underset{\sim}{I}) \delta \underset{\sim}{\phi} \right] dv =$$

$$= - \int \left[\operatorname{div} \underset{\sim}{T} \cdot \delta \underset{\sim}{u} + \operatorname{div} \underset{\sim}{\Lambda} \cdot \delta \underset{\sim}{\phi} + (\underset{\sim}{T} - \underset{\sim}{S}) \delta \phi \right] dv$$

we get the expression for the density of energy w:

$$2w = \lambda (\operatorname{tr} \underset{\sim}{E})^2 + 2\mu \underset{\sim}{E} : \underset{\sim}{E} + 2\tau \operatorname{tr} \underset{\sim}{E} \operatorname{tr} \underset{\sim}{\mathcal{E}} + 4\sigma \underset{\sim}{E} : \underset{\sim}{\mathcal{E}} +$$

$$+ \eta (\operatorname{tr} \underset{\sim}{E})^2 + 2\bar{\nu}\, \underset{\sim}{\mathcal{E}} : \underset{\sim}{\mathcal{E}} + c\, K(\underset{\sim}{\phi})$$

where:

$$K(\underset{\sim}{\phi}) = 2(\phi_{nk,n} \phi_{lk,l} - \phi_{nn,k} \phi_{ll,k})^2 +$$

$$+ \phi_{lk,p}(\phi_{kp,l} - \phi_{kl,p}) + \phi_{kl,p}(\phi_{kp,l} - \phi_{kl,p}) +$$

$$+ \phi_{lp,k}(\phi_{kl,p} - \phi_{kp,l})$$

$K(\underset{\sim}{\phi})$ has not a constant sign therefore we shall henceforth set $c = 0$. The product $\underset{\sim}{E} : \underset{\sim}{\mathcal{E}}$ denotes double contraction. Now the density function w can be presented in the form with the scalar and deviator $(\hat{\underset{\sim}{E}}, \hat{\underset{\sim}{\mathcal{E}}})$ parts separated:

$$2w = (\lambda + \tfrac{2}{3}\mu)(\operatorname{tr} \underset{\sim}{E})^2 + 2\mu \hat{\underset{\sim}{E}} : \hat{\underset{\sim}{E}} + 2(\tau + \tfrac{2}{3}\sigma) \operatorname{tr} \underset{\sim}{E} \operatorname{tr} \underset{\sim}{\mathcal{E}} +$$

$$+ 4\sigma \hat{\underset{\sim}{E}} : \hat{\underset{\sim}{\mathcal{E}}} + (\bar{\eta} + \tfrac{2}{3}\bar{\nu})(\operatorname{tr} \underset{\sim}{\mathcal{E}})^2 + 2\bar{\nu}\, \hat{\underset{\sim}{\mathcal{E}}} : \hat{\underset{\sim}{\mathcal{E}}} .$$

The postulate that w shall be positively defined leads to the following conditions:

$$\left.\begin{aligned}&\lambda+\tfrac{2}{3}\mu\geqslant 0\,,\quad \bar{\eta}+\tfrac{2}{3}\bar{\nu}\geqslant 0\,,\\&/\lambda+\tfrac{2}{3}\mu//\bar{\eta}+\tfrac{2}{3}\bar{\nu}/\geqslant/\tau+\tfrac{2}{3}\sigma/^2\\&\mu\geqslant 0\,,\quad \bar{\nu}\geqslant 0\,,\quad \mu\bar{\nu}\geqslant\sigma^2 \quad\text{and } c=0\end{aligned}\right\}\quad /18/$$

Here we have put for the material constants:
$\bar{\lambda}=\lambda+\eta+\tau$, $\bar{\mu}=\mu+\nu+\sigma$, $\bar{\eta}=\eta-\tau$, $\bar{\nu}=\nu-\sigma$.

In the last part we will discuss the point source solutions with time factor exp $/-i\omega t/$. In these solutions there appear the standing waves. It will be shown that in a special case of choice of the constitutive constants

$$\eta=\bar{\eta}=\tau=0\,,\quad \bar{\lambda}=\lambda \qquad /19/$$

we are avoiding the appearance of the standing waves. So, at last, we can rewrite the strain-stress relations and the expression for energy density function for the symmetric micromorphic continuum - taking into account the above anticipated choice of constitutive constants:

$$\begin{aligned}&T_{nl}=\lambda u_{n,n}\delta_{lk}+2\bar{\mu}u_{(l,k)}+2\nu\phi_{lk}\\&T_{nl}-S_{kl}=-2\nu u_{(l,k)}-2\bar{\nu}\phi_{lk}\\&\Lambda_{plk}=0\\&2w=\lambda E_{ii}E_{ii}+2\mu E_{in}E_{in}+4\sigma E_{in}\varepsilon_{in}+2\bar{\nu}\varepsilon_{in}\varepsilon_{in}\end{aligned}$$

where:

$$\begin{aligned}&\lambda+\tfrac{2}{3}\mu\geqslant 0\,,\quad /\lambda+\tfrac{2}{3}\mu/\bar{\nu}\geqslant\tfrac{2}{3}\sigma^2\,,\\&\mu\geqslant 0\,,\quad \bar{\nu}\geqslant 0\,,\quad \mu\bar{\nu}\geqslant\sigma^2\end{aligned}$$

The equations of motion in terms of displacement $\underset{\sim}{u}$ and microdisplacement $\underset{\sim}{\phi}$ take the form:

$$\left.\begin{aligned}&\lambda u_{n,ni}+\bar{\mu}u_{i,nn}+\bar{\mu}u_{n,in}+\nu\phi_{in,n}+\\&+\nu\phi_{ni,n}+\varrho f_i=-\varrho\ddot{u}_i\\&-\nu u_{i,n}-\nu u_{n,i}-\bar{\nu}\phi_{in}-\bar{\nu}\phi_{ni}+\\&+\varrho L_{ni}=-\varrho I_{kn}\ddot{\phi}_{ik}\end{aligned}\right\}\quad /20/$$

The couples $\underset{\sim}{L}$ appear in the above system parallely with strain $\underset{\sim}{E}$. The body couples are assumed in these equations as the known quantities and according to the above given arguments these couples can be regarded as determined by the external /regional/ stresses governing the deformational processes both in the micro- and macrodeformation scales.

Thus we can put

$$\rho L_{in} = -2\nu E_{in}^{ext}$$

and assuming that for the external field we can apply the usual relations of linear elasticity we will get:

$$\rho L_{in} = -\frac{\nu}{\mu} T_{in}^{ext} \qquad /i \neq n/$$

In the similar way summing the second group of equations /20/ for i = n we get

$$-2\nu E_{ii} - 2\bar{\nu}\phi_{ii} + \rho L_{ii} = -\rho I_{ni}\ddot{\phi}_{in}$$

Now we can put:

$$\frac{2\nu}{3\lambda + 2\mu} p - 2\nu E_{ii}$$

instead of

$$\rho L_{ii} - 2\nu E_{ii}$$

extracting the regional constant pressure from the local part of E_{ii}.

In some applications it is useful to consider the external, regional field through the terms:

$$-\frac{\nu}{\mu} T_{in}^{ext} \quad /i \neq n/ \quad \text{and} \quad \frac{2\nu}{3\lambda + 2\mu} p.$$

7. PLANE PROBLEM AND WAVE DISPERSION

The equations of motion /20/ with $\underset{\sim}{f} = 0$, $\underset{\sim}{L} = 0$ can be groupped into pairs having on the left side the same expressions. By comparison we get /we assume here the same time dependence/

$$\left.\begin{aligned} I_2\phi_{12} &= I_1\phi_{21} \\ I_3\phi_{13} &= I_1\phi_{31} \\ I_3\phi_{23} &= I_2\phi_{32} \end{aligned}\right\} \qquad /21/$$

The above relations can be expressed also in another way:

$$\phi_{[12]} \simeq \phi_{12}/I_{11} - I_{22}/$$
$$\phi_{[31]} \simeq \phi_{31}/I_{33} - I_{11}/$$
$$\phi_{[23]} \simeq \phi_{23}/I_{22} - I_{33}/$$

It means that even the stresses give zero rotation nevertheless the deformations can form microrotation as given by microdisplacements. This explains how the microrotations can be included in the state of deformation especially in the near source zone and this is due to differences in inertia properties of the microelements permeating continuum. Such differences contribute to the rotation of elements as shown by the rotational part of the microdisplacement tensor.

Further on in the plane case we will assume

$$I_1 = I_2 = I \ , \quad I_3 = \alpha I$$

According to relations /21/ we get immediately

$$\phi_{12} = \phi_{21} \ , \quad \phi_{31} = \alpha\,\phi_{13} \ , \quad \phi_{32} = \alpha\,\phi_{23}$$

Now, for the plane waves in form:

$$\exp/imkx_1 + inkx_3 - i\omega t/$$

we get the following two groups of equations /all quantities assummed to be independent of x_2/:

P-group /u_1, u_3, ϕ_{11}, ϕ_{33}, ϕ_{13}/:

$$\left.\begin{aligned}
&- 2\nu u_{1,1} - 2\bar{\nu}\phi_{11} + \varrho\omega^2 I\phi_{11} = 0 \\
&- 2\nu u_{3,3} - 2\bar{\nu}\phi_{33} + \varrho\omega^2 I\alpha\phi_{33} = 0 \\
&- \nu u_{1,3} - \nu u_{3,1} + [\varrho\omega^2 I\alpha - \nu/1 + \alpha/]\,\phi_{13} = 0 \\
&/\lambda + 2\bar{\mu}/u_{1,11} + \bar{\mu}u_{1,33} + \varrho\omega^2 u_1 + /\lambda + \bar{\mu}/u_{3,13} + \\
&+ 2\nu\phi_{11,1} + /1 + \alpha/\nu\phi_{13,3} = 0 \\
&/\lambda + 2\bar{\mu}/u_{3,33} + \bar{\mu}u_{3,11} + \varrho\omega^2 u_3 + /\lambda + \bar{\mu}/u_{1,31} + \\
&+ 2\nu\phi_{33,3} + /1 + \alpha/\nu\phi_{13,1} = 0
\end{aligned}\right\} \quad /22/$$

S-group $/u_2, \phi_{12}, \phi_{23}/$:

$$\left.\begin{aligned}
&-\nu u_{2,1} + /\varrho\omega^2 I - 2\bar{\nu}/\phi_{12} = 0 \\
&-\nu u_{2,3} + [\varrho\omega^2 I\alpha - \bar{\nu}/1 + \alpha/]\phi_{23} = 0 \\
&\bar{\mu}/u_{2,11} + u_{2,33}/ + \varrho\omega^2 u_2 + 2\nu\phi_{12,1} \\
&+ /1 + \alpha/\nu\phi_{23,3} = 0
\end{aligned}\right\} \quad /23/$$

For both groups we will discuss the dispersion characterics as given by the vanishing of the appropriate determinant of the system /22/ or /23/. For the P-group we get the two branches determing wave dispersion:

$$\varrho v^2 = \frac{b \pm /b^2 - 4ac/^{1/2}}{2a} \qquad /24/$$

where:

$$a = - RPS$$

$$b = RPS/M + N/ + RS/1 + \alpha/\nu^2 + 4\nu^2 P/Rm^2 + Sn^2/$$

$$\begin{aligned}
c = &- RPS/\lambda + 2\bar{\mu}/\mu + RS/1 + \alpha/\nu^2[2/\lambda + \bar{\mu}/m^2 n^2 - \\
&- n^2 N - m^2 M] - 4\nu^2 P/Rm^2 N + Sn^2 M/ - \\
&- 4\nu^2[R/1 + \alpha/m^4 + 4Pn^2 m^2 + Sn^4/1 + \alpha/]
\end{aligned}$$

$$M = /\lambda + \bar{\mu}/m^2 + \bar{\mu} \quad , \quad N = /\lambda + \bar{\mu}/n^2 + \bar{\mu}$$

and

$$R = \alpha t - 2\bar{\nu} \, , \quad P = \alpha t - /1 + \alpha/\bar{\nu} \, , \quad S = t - 2\bar{\nu}$$

with $t = I\varrho\omega^2$

The product RPS determines three eventual singularity points in the dispersion curves /24/:

$$I\varrho\omega^2 : \frac{2\bar{\nu}}{\alpha} \, , \quad \frac{/1 + \alpha/\bar{\nu}}{\alpha} \, , \quad 2\bar{\nu} \qquad /25/$$

For the S-group we get only one branch

$$\varrho v^2 = \frac{[2m^2\nu^2 + /\mu + \nu/S]P + n^2\nu^2/1 + \alpha/S}{P\,S}$$

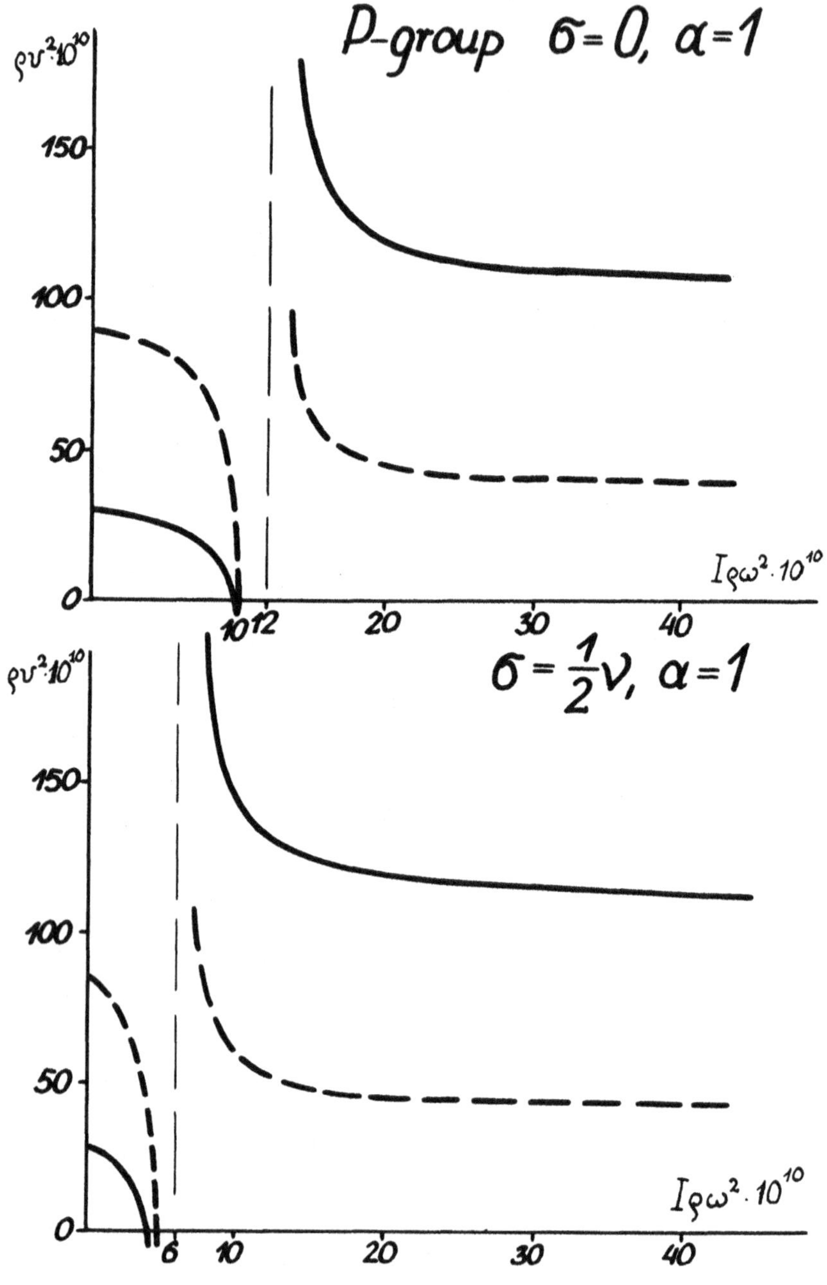

Fig. 1. P-group dispersion: $\alpha = 1.0$, $\sigma = 0$ and $\sigma = \frac{1}{2}\nu$.

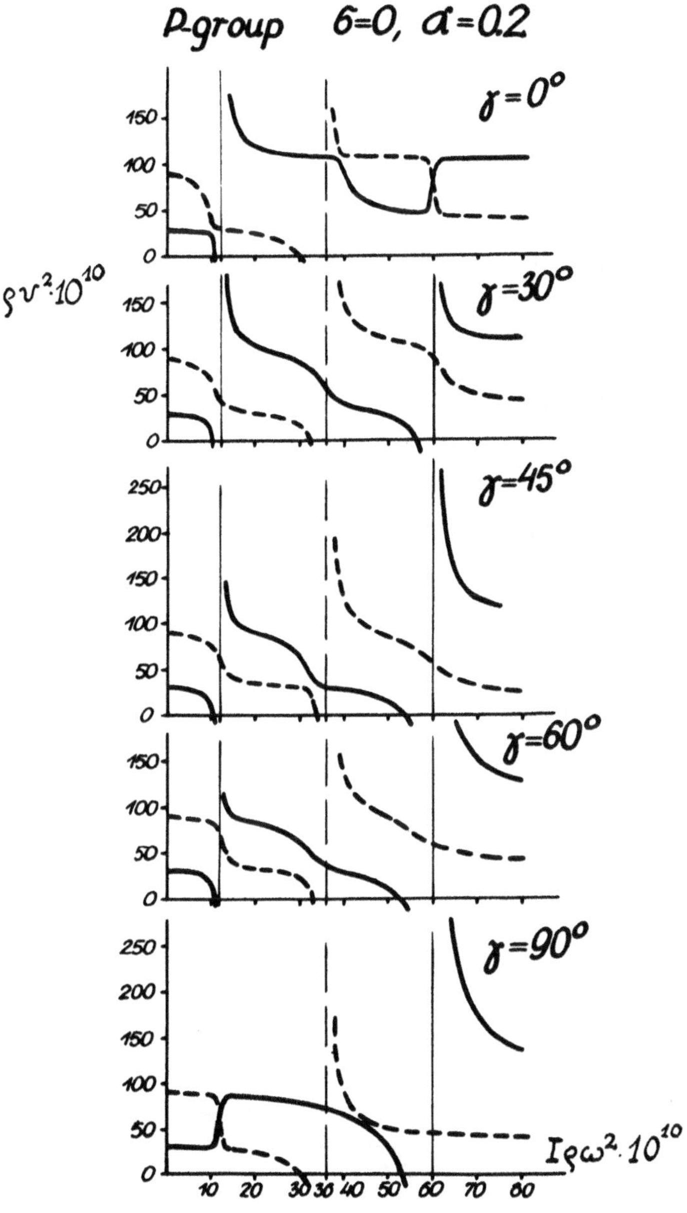

Fig. 2. P-group dispersion: $\alpha = 0.2$, $\sigma = 0$.

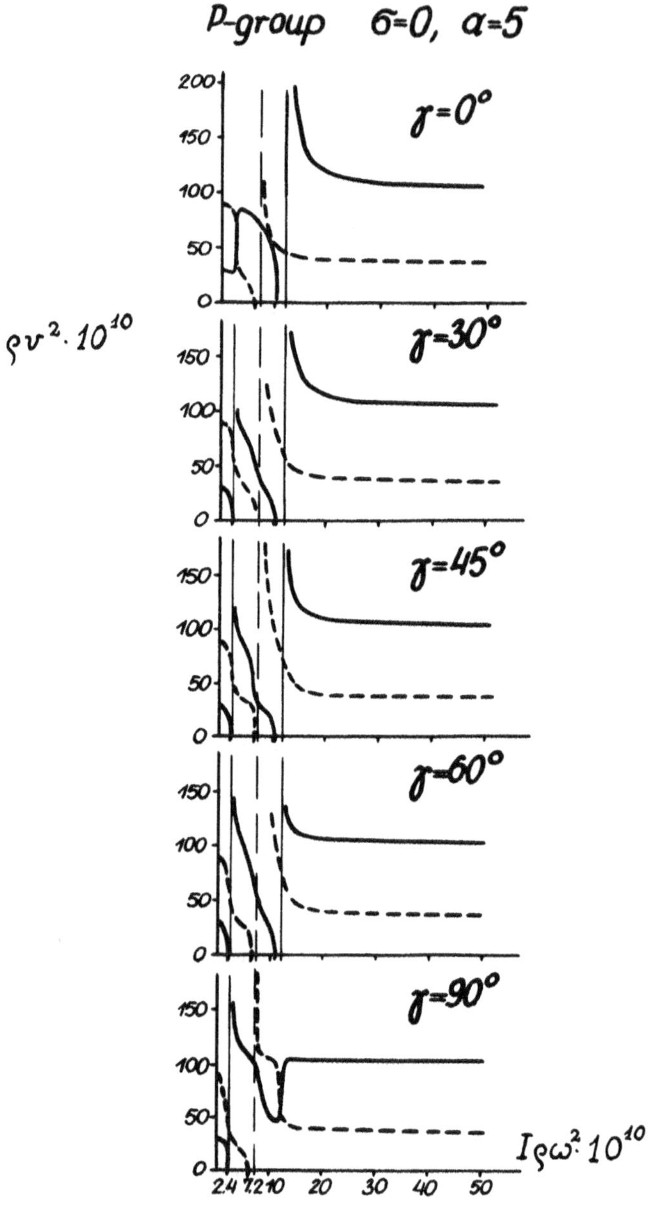

Fig. 3. P-group dispersion: $\alpha = 5.0$, $\sigma = 0$.

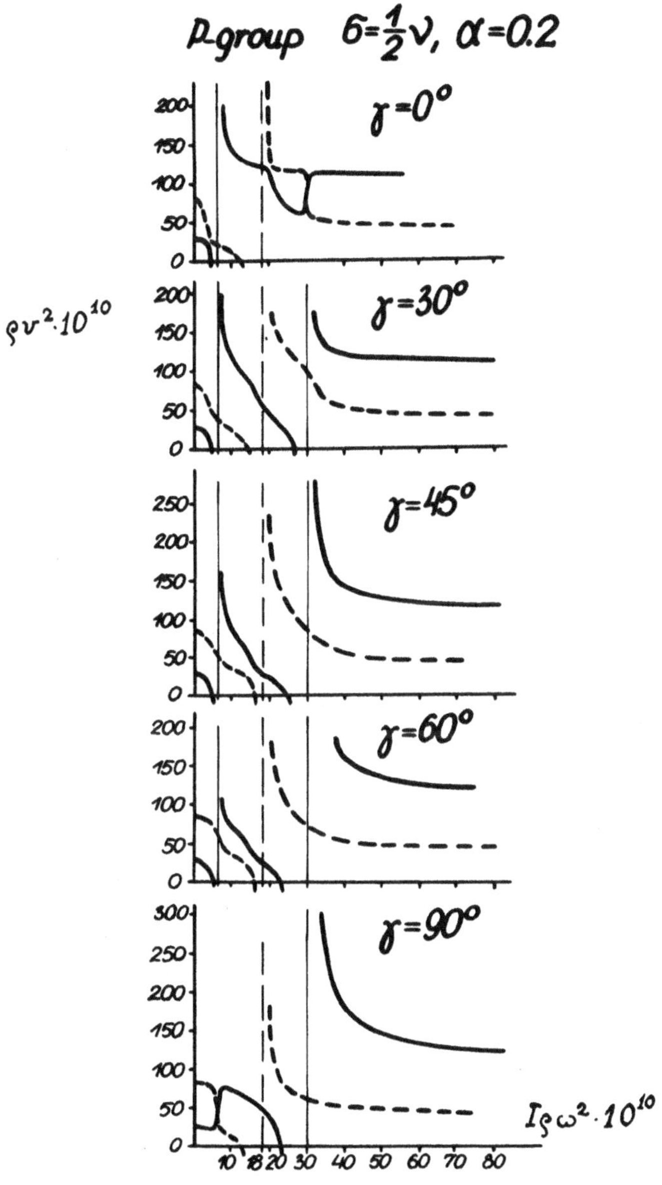

Fig. 4. P-group dispersion: $\alpha = 0.2$, $\sigma = \frac{1}{2}\nu$.

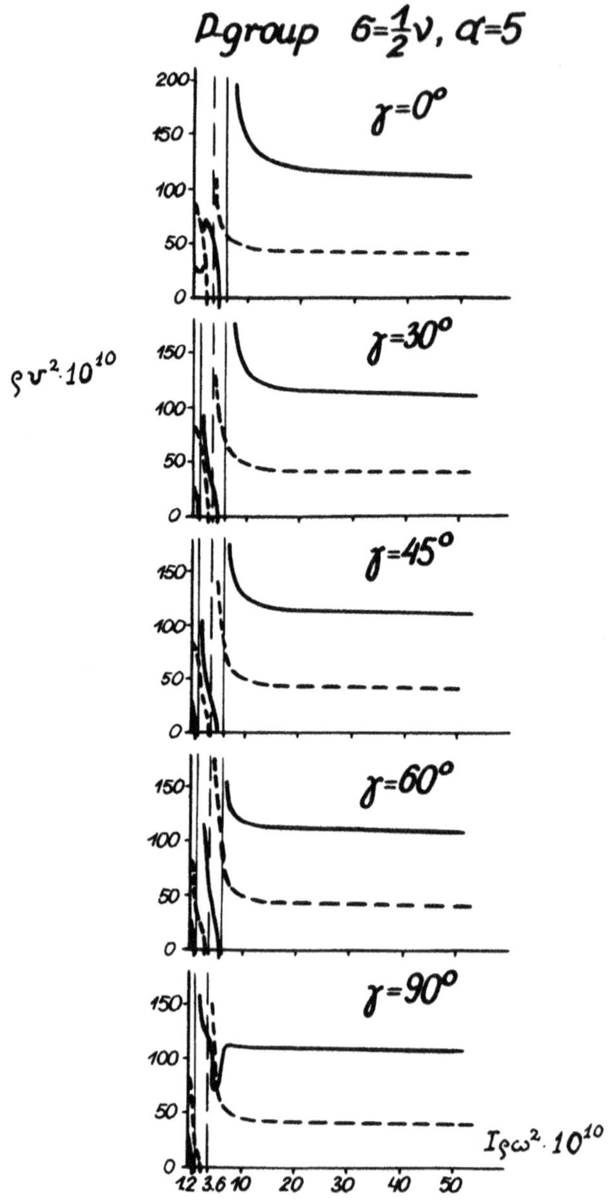

Fig. 5. P-group dispersion: $\alpha = 5.0$, $\sigma = \frac{1}{2}\nu$.

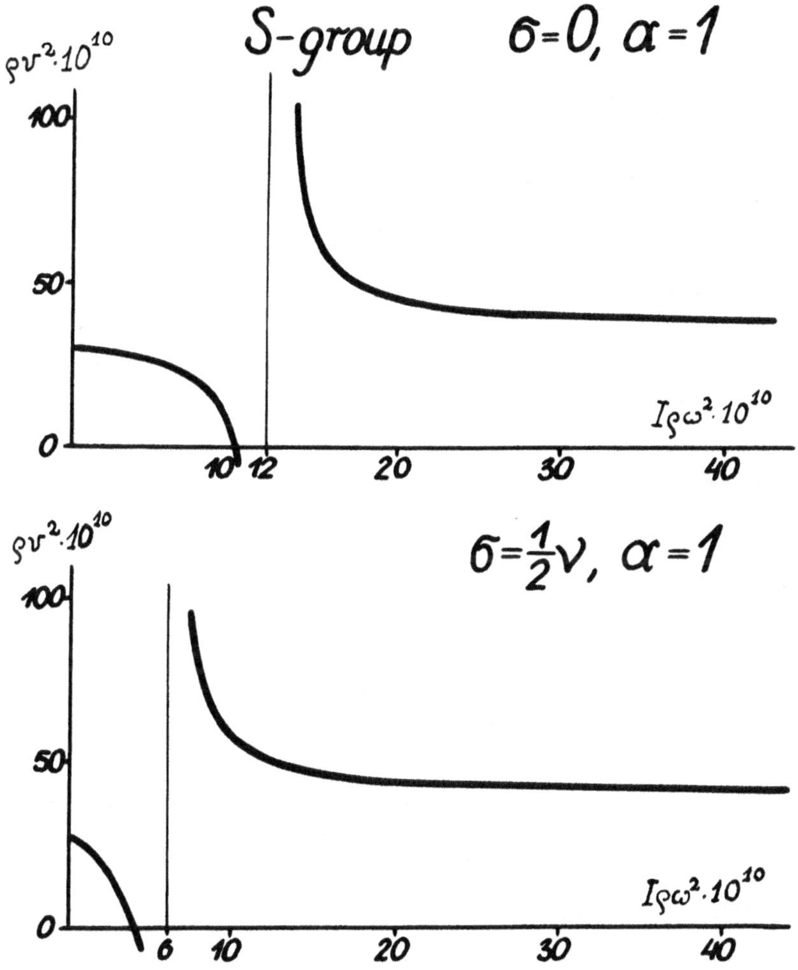

Fig. 6. S-group dispersion: $\alpha = 1.0$, $\sigma = 0$ and $\sigma = \frac{1}{2}$.

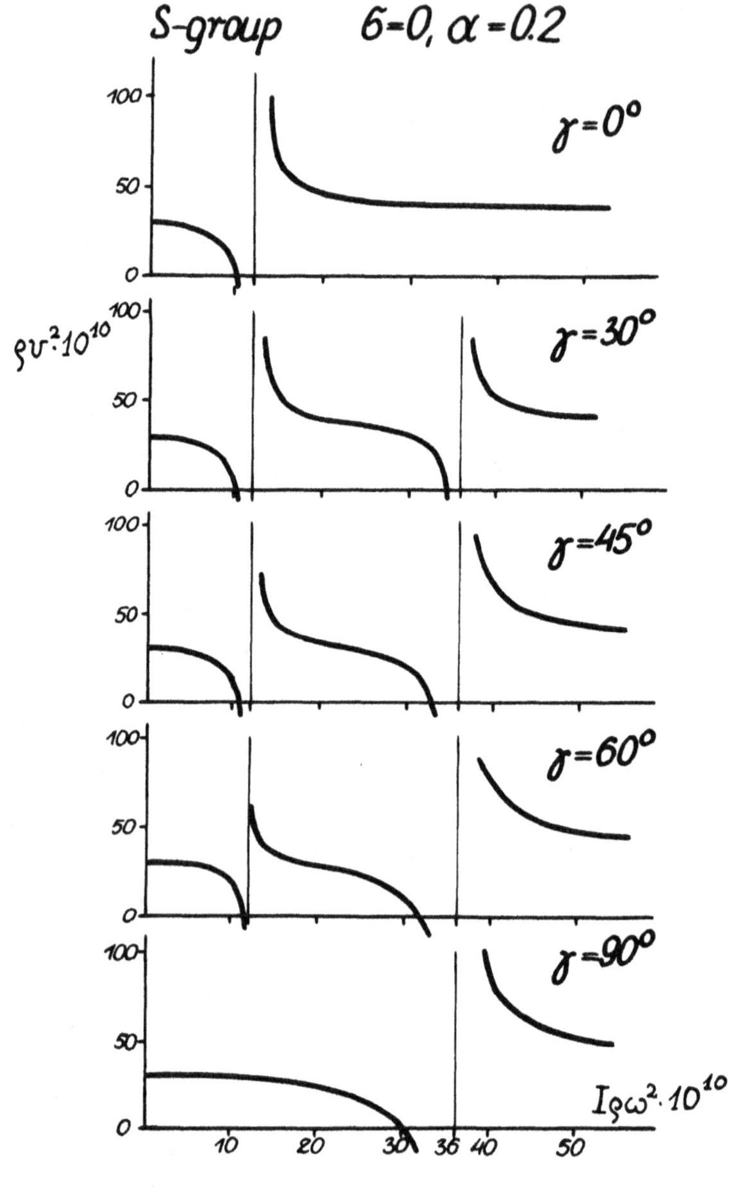

Fig. 7. S-group dispersion: $\alpha = 0.2$, $\sigma = 0$

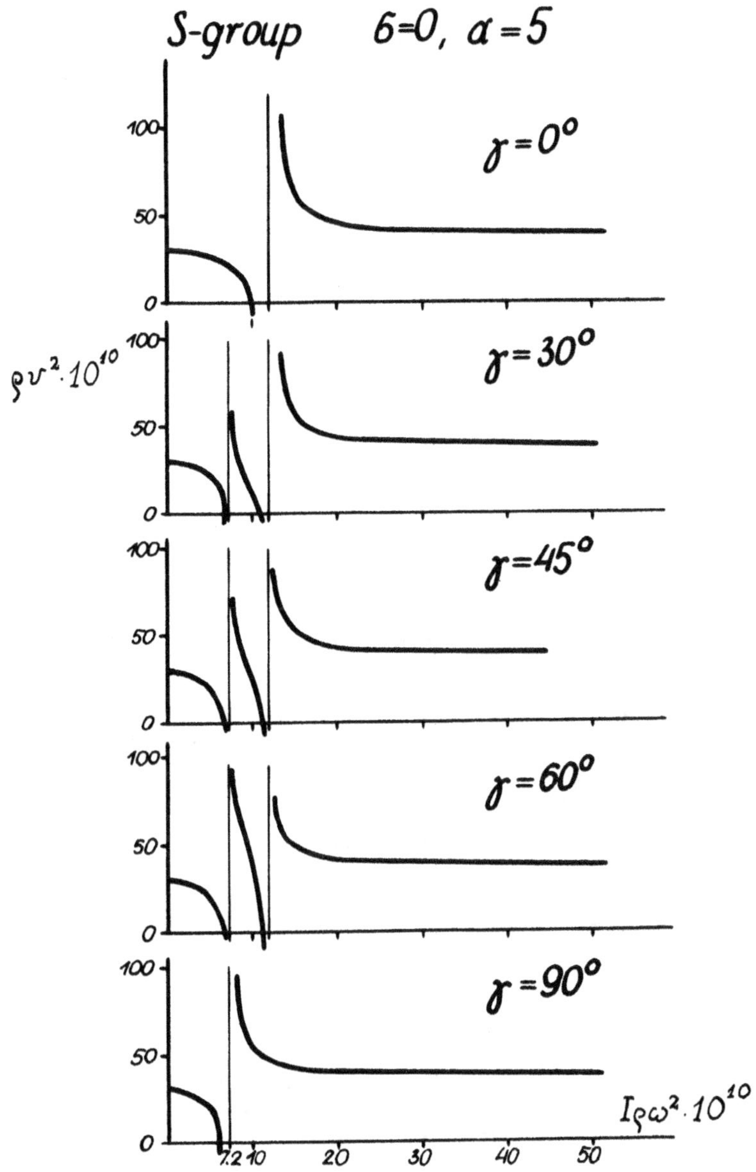

Fig. 8. S-group dispersion: $\alpha = 5.0$, $\sigma = 0$.

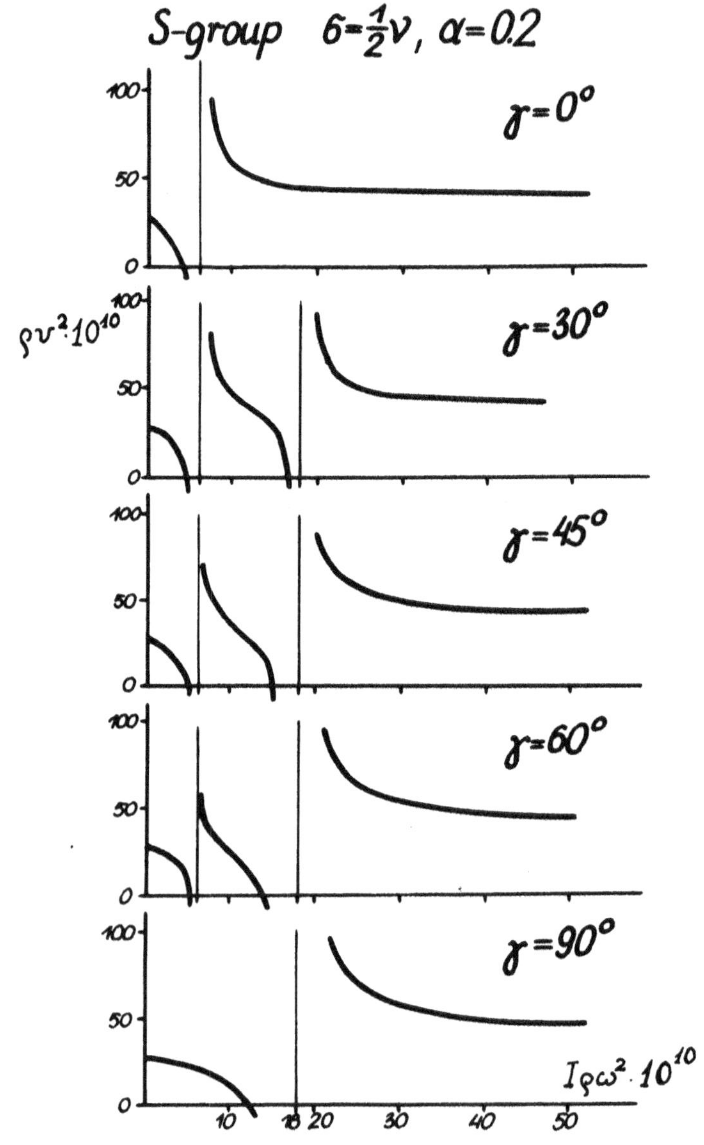

Fig. 9. S-group dispersion: $\alpha = 0.2$, $\sigma = \frac{1}{2}\nu$.

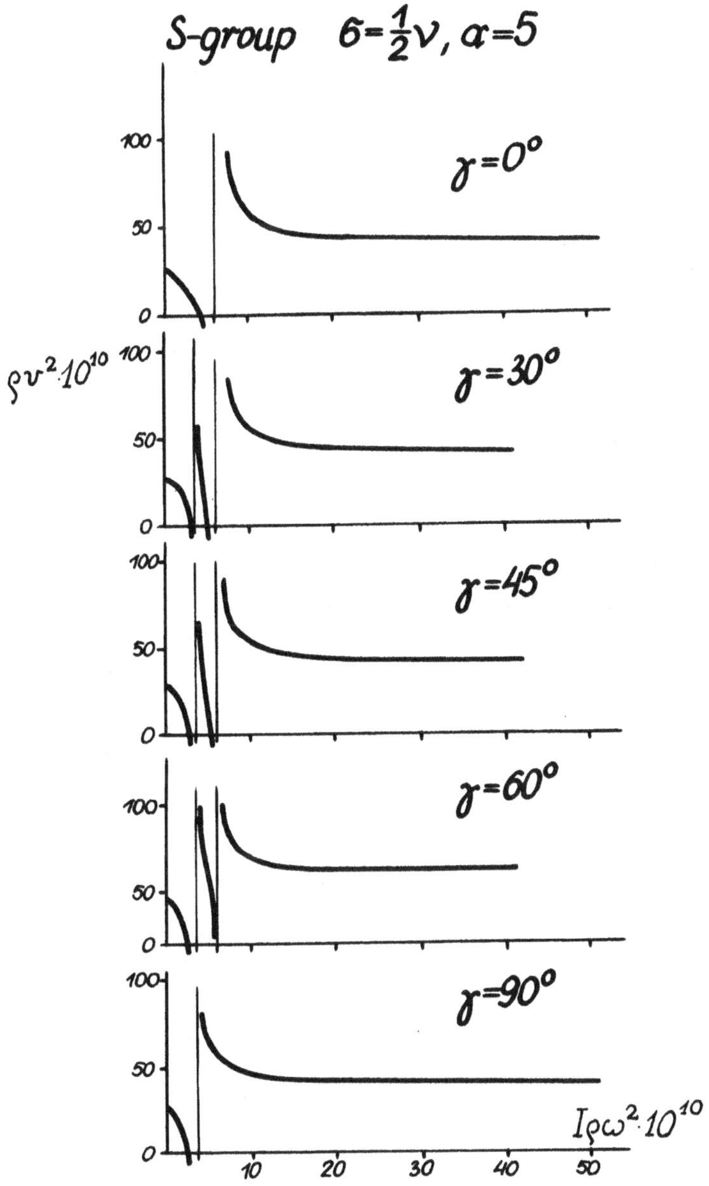

Fig. 10. S-group dispersion: $\alpha = 5.0$, $\sigma = \frac{1}{2}\nu$.

With zeros of the denominator for the values

$$I\varrho\omega^2 : \frac{/1+\alpha/\bar{\nu}}{\alpha}\ , \quad 2\bar{\nu} \qquad /26/$$

The dispersion curves for P- and S-groups are given on the Fig.Fig. 1, 2, 3, 4, 5, 6, 7, 8, 9, 10 for the three values of the inertia ratio α: 1, 0.2, 5 and several directions of wave propagation:

$$\gamma = 0^{\circ}, 30^{\circ}, 45^{\circ}, 60^{\circ}, 90^{\circ} \quad /m = \cos\gamma\ , \quad n = \sin\gamma\ /$$

The propagation in the anisotropy direction $/x_3/$ corresponds to the angle $\gamma = 90^{\circ}$. The choice of parameters is the following $\lambda = \mu$, $\nu = 0.2\mu$ for $\sigma = 0$ and $\sigma = \frac{1}{2}\nu$. The natural length units appear in the micromorphic continua: these units are related to the internal structure. According to /25/ and /26/ the values of those units could be found as related to $I^{1/2}$:

$$\frac{1}{\omega}\left(\frac{2\bar{\nu}}{\alpha\varrho}\right)^{1/2}\ , \quad \frac{1}{\omega}\left(\frac{/1+\alpha/\bar{\nu}}{\alpha\varrho}\right)^{1/2}\ , \quad \frac{1}{\omega}\left(\frac{2\bar{\nu}}{\varrho}\right)^{1/2} \qquad /27/$$

The presented curves show the influence of the medium anisotropy on characteristics of wave dispersion.

8. ROTATIONAL EFFECTS AT THE EPICENTER ZONES

Let us remind that residual rotational deformations were frequently observed in the epicentral areas. It is worth to note that twisting of columns and monuments could be related to rotational waves in the micromorphic continuum.

W.W. Hobbs /1907/ described twisting of obelisks in Calabria /earthquake of 1738/ and has underlined that these rotational effects take place only in some favourable places. The observations in Calabria seems very interesting as obelisks were twisted stepwisely in three vertical sections. Describing the same earthquake R. Mallet has suggested that an in dependent vorticose movement probably took place /Davison, 1927/.

Similar effects have been observed in cases of even not very strong earthquakes, for example: Rhine 1878. Another observations relate to the following earthquakes: Kingstone /1907/, India /1897/, Belluno /1873/. A. Imamura /1937/ discussed the observations by Milne, stating that deformations of twist type occur very often. Twisting of short columns of rectangular section has been explained by Imamura as related to rotational moment which arises when column is tilted by seismic vibrations; especially this effect can appear when seismic impulse cause tilting of column in the direction close to its diagonal: edge of column form the axis of rotation while direction of twisting depends on inertia moments of a column. The different twist directions were noticed by Milne /Davison, 1927/, on contrary in Yokohama /1880/ all grave monuments have been twisted in the same sense. It seems that rotational moments caused by tilting of columns can not explain all observed cases; columns of nearly circular sections would not twist and moreover stepwise twisting in the consecutive vertical sections would be not probable. Thus it seems that at least in some cases the described effects are related to rotational waves which could propagate due to specific properties of medium. B. Gutenberg wrote in his book /Grundlagen der Erdbedenkunde, 1926/ that concerning "drehbewungen des Bodens ... dass sie sichtbar werden, gar nicht denkbar".

In the symmetric micromorphic continuum /Teisseyre, 1973/ the rotational effects are related to microinertia tensor which describes inertia moments of blocks forming microstructure - eq. /21/.

Apparently the explanation of Imamura related to twisting of short columns differs from the present approach. Let us, however, note that rotational effects related to inertia moments of internal blocks reminds inertia effects of columns located on ground surface.

Now, we will discuss the possibility of recording the rotational deformations. This problem has appeared to us at the detailed determinations of azimuths by means of the so called azimuth system of six horizontal seismographs /seismograph positions differing by the 30° degrees/. The observations concern the hypocentral zone; seismographs were located very closely to it.

To interprete the rotational effects on seismograms let us discuss the equation of motion of seismograph in the more general case, when ground motion is determined by displacement X, tilting ψ and twisting γ :

$$\ddot{\Theta} + 2\varepsilon_1 \dot{\Theta} + \eta_1^2 \Theta = \frac{\ddot{X}}{l_o} + \ddot{\gamma} + \frac{g}{l_o}\psi$$

for mechanical system, and

$$\ddot{\Theta} + 2\varepsilon_1 \dot{\Theta} + \eta_1^2 \Theta = \frac{\ddot{X}}{l_o} + \ddot{\gamma} + \frac{g}{l_o}\psi + k\frac{K_2}{K_1}\dot{\varphi}$$

$$\ddot{\varphi} + 2\varepsilon_2 \dot{\varphi} + \eta_2^2 \varphi = k\dot{\Theta}$$

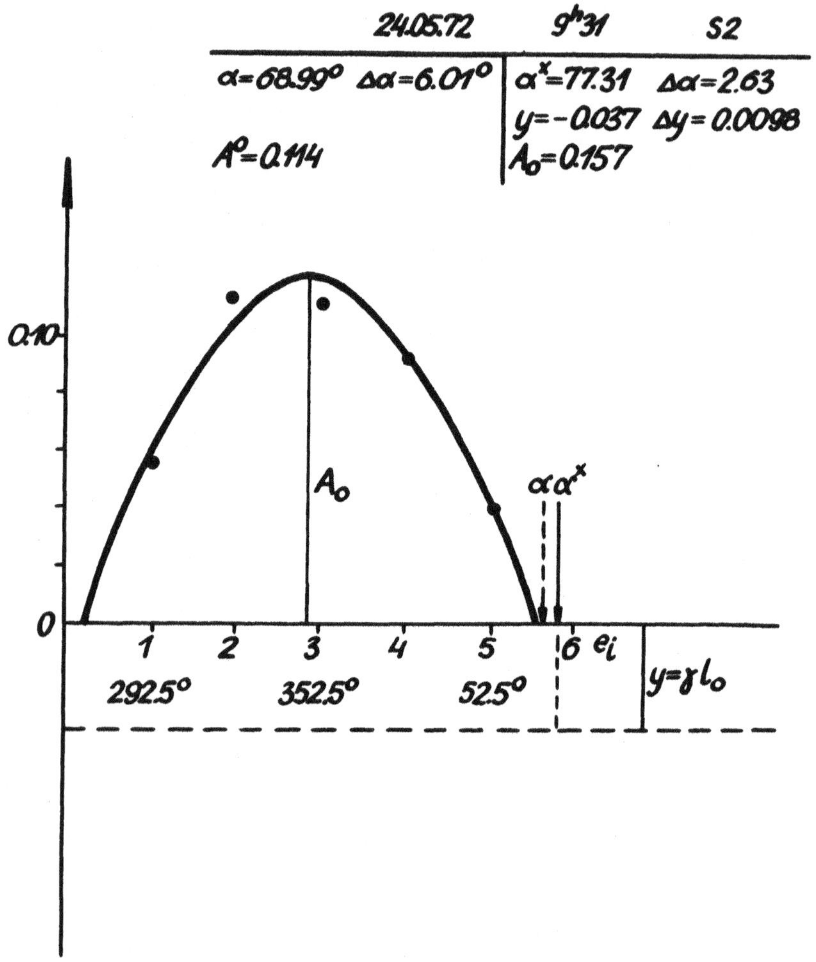

Fig. 11. Azimuth determination and the effect of rotational shifting.

for galvanometric system /usual notation/.
In the case $\psi = 0$ /no tilting/ we get for steady motion

$$y = \frac{2A}{l_o}\left(\frac{K_1}{K_2}\right)^{1/2}\left(\frac{D_1}{D_2}\frac{T_2}{T_1}\sigma^2\right)^{1/2} W/x - l_o\gamma/$$

or $y = V_o\ W/x - l_o\gamma/$.
It follows that we record the value $x - l_o\gamma$ / γ is here defined as positive in clockwise sense/. Thus in the case of the azimuth system of seismographs the maximal amplitudes of P wave recorded by the seismograph with index i will be given as follows

$$A_i^{max} = A_o \sin/e_i + e/ + l_o\gamma$$

where: A_o ground motion amplitude of P wave
e_i - azimuth of the seismograph
e - azimuth of P wave impulse
l_o - reduced length of seismograph
γ - twisting of ground
For the system of seismographs /with constant l_o/ we get the displacement l_o as shown on Fig. 11. Reversely it appears possible to determine the value of rotational deformation γ tracing the shift y of amplitudes of P wave recorded by azimuthal system:

$$\gamma = \frac{y}{l_o}$$

Basing on the given above consideration we will discuss some recording data gained by the azimuth system. The amplitudes of P wave were used to determine the azimuth of wave approach. We could applied two numerical programs. In the first program the standard method of azimuth determination is applied. Second program takes into account the possibility of shifting the reference line /Fig. 11/ as due to rotational deformation. In the most cases /total number of cases under investigation - 150 /the shifts of reference line are small and in the limits of errors. However, for several cases /about 10 %/ the shifts of reference line become remarkable, their erroes being comparatively small. All measurements relate to near hypocentral regions. On Fig. 11 there is given one of several examples for which the appearance of rotational wave seems very probable. It shall be noted that in the discussed cases the determination of azimuths bring different values as comparing with determinations by standard method. Moreover, the errors of azimuth determination are in the discussed cases much smaller when allowing shifts of reference line. Reversely, when the rotational effects seem rather doubtful, the determinations of azimuths by the standard method are accompanied with smaller errors.

The mentioned above method enables to construct a whole seismogram of rotational component as shift residua of the consequtive calculations of azimuths. On Fig. 12 an example presents only those values of the zero line shifts, which are accompanied by distinctly small errors. Times of the P- and S-wave emergences are indicated by arrows. Epicentral distance is here extremaly small. Rotational wave effects are very clear close to the S-wave arrival.

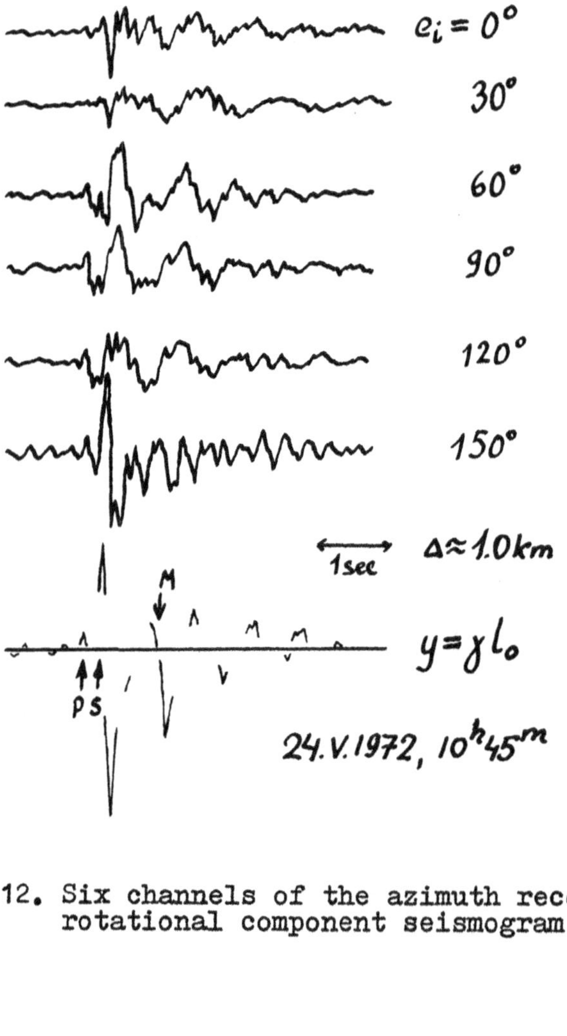

Fig. 12. Six channels of the azimuth record and the rotational component seismogram.

9. POTENTIALS IN THE SYMMETRIC MICROMORPHIC THEORY

Further on we will assume that:

$$I_{lk} = I_l \delta_{lk}$$

and that the principal moments are constant /Teisseyre, 1973/. Taking the time factor in the form exp /$-i\omega t$/ we shall rewrite eqs /20/ in the vector notation:

$$\begin{aligned} &\operatorname{div} \underset{\sim}{T} - \varrho \ddot{\underset{\sim}{u}} = -\varrho \underset{\sim}{f} \\ &\operatorname{div} \underset{\sim}{\Lambda} + /\underset{\sim}{T} - \underset{\sim}{S}/ - \varrho \ddot{\underset{\sim}{\phi}} \underset{\sim}{I} = -\varrho \underset{\sim}{L} \end{aligned} \qquad /28/$$

Further on we will assume the symmetry of the microdisplacement tensor which immediately requiries the equality of the principipal inertia moments:

$$\phi_{lk} = \phi_{kl}, \quad I_1 = I_2 = I_3 = I \qquad /29/$$

The system /28/ can be transform to a convenient form involving only scalar differential operators by means of the following potential functions /Hanyga, Teisseyre, 1974/:
microdisplacements:

$$\begin{aligned} \underset{\sim}{\phi} = \Delta \varkappa \underset{\sim}{1} &+ \operatorname{grad} \operatorname{grad} \varphi + \operatorname{grad}_1 \operatorname{curl}_2 \underset{\sim}{\Psi} \\ &+ \operatorname{grad}_2 \operatorname{curl}_1 \underset{\sim}{\Psi} + \operatorname{curl}_1 \operatorname{curl}_2 \underset{\approx}{\Xi}, \end{aligned} \qquad /30/$$

displacements:

$$\underset{\sim}{u} = \operatorname{grad} \xi + \operatorname{curl} \underset{\sim}{\eta} \; , \qquad /31/$$

body forces:

$$\underset{\sim}{f} = \operatorname{grad} g + \operatorname{curl} \underset{\sim}{h} \; , \qquad /32/$$

body couples:

$$\begin{aligned} \underset{\sim}{L} = \Delta l \underset{\sim}{1} &+ \operatorname{grad} \operatorname{grad} m + \operatorname{grad}_1 \operatorname{curl}_2 \underset{\sim}{n} \\ &+ \operatorname{grad}_2 \operatorname{curl}_1 \underset{\sim}{n} + \operatorname{curl}_1 \operatorname{curl}_2 \underset{\sim}{M} \end{aligned} \qquad /33/$$

The following auxillary conditions can be imposed on the potential functions:

$$\left.\begin{aligned} &\operatorname{div} \underset{\sim}{\Psi} = 0, \quad \operatorname{div} \underset{\approx}{\Xi} = 0, \quad \operatorname{tr} \underset{\approx}{\Xi} = 0, \quad \operatorname{div} \underset{\sim}{\eta} = 0 \\ &\operatorname{div} \underset{\sim}{n} = 0, \quad \operatorname{div} \underset{\sim}{M} = 0, \quad \operatorname{tr} \underset{\sim}{M} = 0, \quad \operatorname{div} \underset{\sim}{h} = 0 \end{aligned}\right\} /34/$$

The index 1 or 2 after differential operator indicates here above the first or second tensor index to which it refers. We shall also note some interesting differential relations:

$$\operatorname{curl}_1 \operatorname{curl}_2 \underset{\sim}{\phi} = \Delta^2 \underset{\sim}{\Xi}, \quad \operatorname{tr} \underset{\sim}{\phi} = \Delta\varphi + 3\Delta\varkappa,$$

$$\operatorname{curl}_1 \operatorname{div}_2 \underset{\sim}{\phi} = \operatorname{curl}_2 \operatorname{div}_1 \underset{\sim}{\phi} = -\Delta^2 \underset{\sim}{\psi},$$

$$\operatorname{div}_1 \operatorname{div}_2 \underset{\sim}{\phi} = \Delta^2 \varkappa + \Delta^2 \varphi$$

The constitutive relations we will take now in the form given by eqs. /15/ with conditions /16-18/. As we have mentioned before condition /19/ has been taken as anticipated one; thus we begin with the case of non vanishing parameters $\eta, \bar{\eta}, \tau$.

Applying now the suitable differential operators to eqs. /28/ we can easily express the system of equations in terms of potential /30-33/:

P-type equations:

$$\left.\begin{aligned} &/\bar{\lambda} + 2\bar{\mu}/\Delta\xi + /\eta + 2\nu/\Delta\varphi + \\ &+ /3\eta + 2\nu/\Delta\varkappa + \varrho\omega^2\xi = -\varrho g, \\ &-/3\eta + 2\nu/\Delta\xi + /\varrho\omega^2 I - 3\bar{\eta} - 2\bar{\nu}//3\Delta\varkappa + \Delta\varphi/ = \\ &= -3\varrho l - \varrho\Delta m, \\ &-/\eta + 2\nu/\Delta\xi + /\varrho\omega^2 I - 3\bar{\eta} - 2\bar{\nu}/\Delta\varkappa + \\ &+ /\varrho\omega^2 I - \bar{\eta} - 2\bar{\nu}/\Delta\varphi = -\varrho l - \varrho\Delta m; \end{aligned}\right\} \quad /35/$$

S-type equations:

$$\left.\begin{aligned} &\bar{\mu}\Delta\underset{\sim}{\eta} + 2\nu\Delta\underset{\sim}{\psi} + \varrho\omega^2\underset{\sim}{\eta} = -\varrho\underset{\sim}{h}, \\ &\nu\Delta\underset{\sim}{\eta} - /\varrho\omega^2 I - 2\bar{\nu}/\Delta\underset{\sim}{\psi} = \varrho\Delta\underset{\sim}{n}; \end{aligned}\right\} \quad /36/$$

M-type equations:

$$/\varrho\omega^2 I - 2\bar{\nu}/\underset{\approx}{\Xi} = -\varrho\underset{\sim}{M}.$$

The last equation is independent and only related to stress moments. For the source model in our symmetric micromorphic theory we can put $\underset{\sim}{M} = 0$ and $\underset{\approx}{\Xi} = 0$.

10. POINT SOURCE SOLUTIONS

First, to solve the systems of P- and S-type for a point source we shall seek the expressions for potentials due to body force and body couple concentrated in a point. The point force or couple is defined as the limit of the respective density over a volume which contracts to a point:

$$\bar{f} = \lim_{v\to 0}\int \rho \underset{\sim}{f}\, dv\ , \qquad \underset{\sim}{\bar{L}} = \lim_{v\to 0}\int \rho \underset{\sim}{L}\, dv\ .$$

Now, from the definition /32/ we get

$$g_{,ll} = f_{l,l}\ , \qquad h_{p,ll} = -\, e_{prl} f_{l,r}$$

and for a point source $/x_o/$ the following expressions for potentials g, h are obtained

$$\left.\begin{aligned} g &= -\frac{1}{4\pi\rho}\,\bar{f}_l\,\frac{\partial}{\partial x_l}\,\frac{1}{r}\ ,\\ h_p &= \frac{1}{4\pi\rho}\,e_{prl}\bar{f}_l\,\frac{\partial}{\partial x_r}\,\frac{1}{r}\ , \end{aligned}\right\} \qquad /37/$$

where: $r/x - x_o/$ is the distance from the source.

In a similar way we can get from the definitions /33/ with help of the auxiliary conditions /34/

$$\Delta l = -\frac{1}{2}L_{lk,lk}\ , \qquad m_{,ll} = -\,3\Delta l\ ,$$

$$n_{p,llrr} = -\, e_{prl}L_{lk,rk}$$

Thus, for a point source we have

$$\left.\begin{aligned} \Delta l &= -\frac{1}{8\pi\rho}\,\bar{L}_{lk}\,\frac{\partial}{\partial x_l\,\partial x_k}\,\frac{1}{r}\ ,\\ n_{p,rr} &= -\frac{1}{4\pi\rho}\,e_{prl}\bar{L}_{lk}\,\frac{\partial}{\partial x_r\,\partial x_k}\,\frac{1}{r} \end{aligned}\right\} \qquad /38/$$

Now, we can seek the formal solution of the P-system: eqs. /35/ and of the S-system: eqs. /36/. After some rearrangement of the eqs. /35/ we get the following system:

$$\left.\begin{aligned} &H\Delta\varkappa + N\Delta\xi + \varrho\omega^2\bar{\eta}\xi = -\bar{\eta}\varrho g - /\eta + 2\nu/\varrho\Delta l \ , \\ &K\Delta\varkappa + M\Delta\xi + \varrho\omega^2 P\xi = -P\varrho g - 2/\eta + 2\nu/\varrho\Delta l \ , \\ &2\nu\Delta\xi - S\Delta\varphi = -3\varrho\Delta l \ , \end{aligned}\right\} /39/$$

We introduce the notation:

$$\left.\begin{aligned} &H = \bar{\eta}/3\eta + 2\nu/ + /\eta + 2\nu/T \ , \\ &N = \bar{\eta}/\bar{\lambda} + 2\bar{\mu}/ - \eta/\eta + 2\nu/ \ , \\ &K = 2\eta S + 4\bar{\eta}\nu \ , \quad M = /\bar{\lambda} + 2\bar{\mu}/P + /\eta + 2\nu/^2 \\ &S = \varrho\omega^2 I - 2\bar{\nu} \ , \quad P = \varrho\omega^2 I - \bar{\eta} - 2\bar{\nu} , \\ &T = \varrho\omega^2 I - 3\bar{\eta} - 2\bar{\nu} , \\ &W = KN - HM \end{aligned}\right\} /40/$$

To solve the system /39/ we can apply the Fourier transform and thus after somewhat lengthy calculations we can get the Green functions for the potentials $\varkappa$, ξ, φ. For the Fourier transforms $\tilde{\varkappa}$, $\tilde{\xi}$, $\tilde{\varphi}$ we get respectively:

$$\left.\begin{aligned} &\tilde{\varkappa} = \frac{-1}{2k^2k_0W}\left(\frac{1}{k-k_0} - \frac{1}{k+k_0}\right)\varrho g\big[\bar{\eta}k^2/M - N/ \\ &- k^2NS\big] + \varrho\Delta l/\eta + 2\nu/\big[k^2/M + 2N/ - \varrho\omega^2 P \ , \end{aligned}\right\} /41/$$

$$\left.\begin{aligned} &\tilde{\xi} = \frac{1}{W}\,\frac{1}{2k_0}\left(\frac{1}{k-k_0} - \frac{1}{k+k_0}\right)\Big\{\varrho g\big[\bar{\eta}/K + H/ - HS\big] \\ &+ \varrho\Delta l/\eta + 2\nu//K - 2H/\Big\} \ , \end{aligned}\right\} /42/$$

$$\left.\begin{aligned} &\tilde{\varphi} = \frac{2\nu}{SW}\,\frac{1}{2k_0}\left(\frac{1}{k-k_0} - \frac{1}{k+k_0}\right)\Big\{\varrho g\big[\bar{\eta}/K + H/ - HS\big] \\ &+ \varrho\Delta l/\eta + 2\nu//K - 2H/ \ - \frac{3}{S}\,\frac{1}{k^2}\,\varrho\Delta l \ , \end{aligned}\right\} /43/$$

where:

$$k_0^2 = \frac{\varrho\omega^2\bar{\eta}K - \varrho\omega^2 HP}{W} = \frac{-\varrho\omega^2 ST/\eta + 2\nu/}{W} \ .$$

The body forces and couples are given here above by potentials ϱg and $\varrho\Delta l$, thus we can calculate the respective Green function denoting by G the part acting on ϱg and by $\bar{G}$ that related to $\varrho\Delta l$.

$$\left.\begin{aligned} G_{\varkappa} &= \frac{1}{2\pi W}\left[\eta/M - N/ - NS\right]\frac{\exp/ik_oR/}{R}, \\ \bar{G}_{\varkappa} &= \frac{1}{2\pi k_o^2 W}/\eta + 2\nu/\left[k_o^2/M - 2N/ - \varrho\omega^2 P\right]\cdot \\ &\frac{\exp/ik_oR/}{R}, \end{aligned}\right\} \quad /44/$$

$$\left.\begin{aligned} G_{\xi} &= -\frac{1}{2\pi W}\left[\bar{\eta}/K + H/ - HS\right]\frac{\exp/ik_oR/}{R} \\ \bar{G}_{\xi} &= -\frac{1}{2\pi W}/\eta + 2\nu//K - 2H/\ \frac{\exp/ik_oR/}{R} \end{aligned}\right\} \quad /45/$$

$$\left.\begin{aligned} G_{\varphi} &= -\frac{\nu}{\pi SW}\left[\bar{\eta}/K + H/ - HS\right]\frac{\exp/ik_oR/}{R} \\ \bar{G}_{\varphi} &= -\frac{\nu}{\pi SW}/\eta + 2\nu//K - 2H/\ \frac{\exp/ik_oR/}{R} + \\ &+ \frac{3}{2\pi S}\ \frac{1}{R} \end{aligned}\right\} \quad /46/$$

where: $R^2 = /x_1' - x_1//x_1' - x_1/.$

For the last term the integration contour related to the residuum $k = 0$ is shown on Fig. 13. Now, we can easily calculate the potentials $\varkappa$, ξ, φ according to the formula, as given here beneath for $\varkappa$:

$$\begin{aligned} \varkappa/x,x_o/ &= \int G/x',x/\varrho\ g/x',x_o/d_{(3)}x' + \\ &+ \int \bar{G}/x',x/\varrho\Delta l/x',x_o/d_{(3)}x', \end{aligned} \quad /47/$$

We get:

$$\left.\begin{aligned} \varkappa &= \frac{-1}{2\pi}\bar{f}_1\left(\frac{\partial}{\partial x_1}\ \frac{1}{R}\right)\frac{\bar{\eta}/M - N/ - NS}{k_o^2 W}\ \cdot \\ &\cdot\left[-ik_oR_o\exp/ik_oR_o/ + \exp/ik_oR_o/ - 1\right] \\ &- \frac{1}{4\pi}\bar{L}_{1k}\left(\frac{\partial^2}{\partial x_1\partial x_k}\ \frac{1}{R_o}\right)\frac{/\eta + 2\nu/\left[k_o^2/M - 2N/ - \varrho\omega^2 P\right]}{k_o^4}\cdot \\ &\cdot\left[-ik_oR_o\exp/ik_oR_o/ + \exp/ik_oR_o/ - 1\right], \end{aligned}\right\} \quad /48/$$

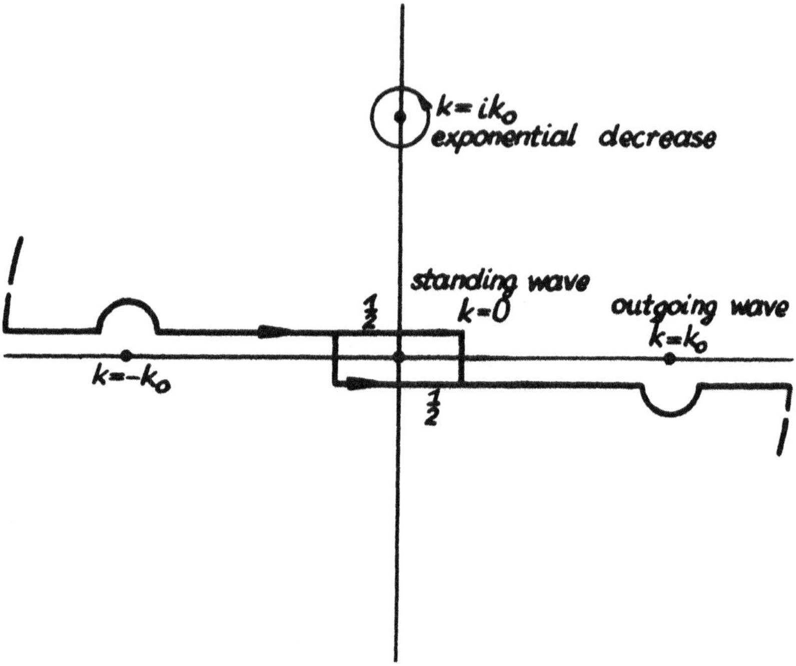

Fig. 13. Integration contour.

$$
\begin{aligned}
\xi = & - \frac{1}{2\pi} \bar{f}_1 \left(\frac{\partial}{\partial x_1} \frac{1}{R_o} \right) \frac{/\eta + 2\nu/ST}{k_o^2\, W} \cdot \\
& \left[-ik_oR_o \exp/ik_oR_o/ + \exp/ik_oR_o/ - 1 \right] \\
& - \frac{\nu}{\pi} \bar{L}_{1k} \left(\frac{\partial^2}{\partial x_1 \partial x_k} \frac{1}{R_o} \right) \frac{/\eta + 2\nu/T}{k_o^2\, W} \cdot \\
& \left[-ik_oR_o \exp/ik_oR_o/ + \exp/ik_oR_o/ - 1 \right],
\end{aligned}
\qquad /49/
$$

$$
\begin{aligned}
\varphi = & - \frac{\nu}{\pi} \bar{f}_1 \left(\frac{\partial}{\partial x_1} \frac{1}{R_o} \right) \frac{/\eta + 2\nu/ST}{k_o^2 SW} \cdot \\
& \left[-ik_oR_o \exp/ik_oR_o/ + \exp/ik_oR_o/ - 1 \right] \\
& - \frac{2\nu}{\pi} \bar{L}_{1k} \left(\frac{\partial^2}{\partial x_1 \partial x_k} \frac{1}{R_o} \right) \frac{\nu T/\eta + 2\nu/}{k_o^2 SW} \cdot
\end{aligned}
\qquad /50/
$$

$$\left.\begin{aligned}&\cdot\left[-ik_oR_o\exp/ik_oR_o/+\exp/ik_oR_o/-1\right]\\&-\frac{3}{8\pi}\bar{L}_{lk}\left(\frac{\partial^2}{\partial x_l\partial x_k}\frac{1}{R_o}\right)\frac{1}{S}R_o^2\end{aligned}\right\}\quad /50/$$

where: $R_o^2 = /x_l - x_{ol}//x_l - x_{ol}/$.

For the S-system /36/ the similar calculations bring for the Fourier transforms $\tilde{\eta}_l$, $\tilde{\psi}_l$:

$$\left.\begin{aligned}\tilde{\eta}_l &= \frac{1}{2\bar{k}_o}\left(\frac{1}{k-\bar{k}_o}-\frac{1}{k+\bar{k}_o}\right)\cdot\\&\left[\rho h_l\frac{\bar{k}_o}{\rho\omega^2}-\rho n_{l,rr}\frac{2\nu}{\bar{\mu}S+2\nu^2}\right],\end{aligned}\right\}\quad /51/$$

$$\left.\begin{aligned}\tilde{\psi}_l &= \frac{1}{2\bar{k}_o}\left(\frac{1}{k-\bar{k}_o}-\frac{1}{k+\bar{k}_o}\right)\cdot\\&\left[\rho h_l\frac{\nu}{\bar{\mu}S+2\nu^2}-\rho n_{l,rr}\frac{/\rho\omega^2-\bar{\mu}k_o^2/}{k^2/\bar{\mu}S+2\nu^2/}\right]\\&+\rho n_{l,rr}\frac{1}{k^2}\frac{\rho\omega^2}{S},\end{aligned}\right\}\quad /52/$$

where: $\bar{k}_o^2 = \dfrac{\rho\omega^2 S}{\bar{\mu}S + 2\nu^2}$.

Here above the body forces and couples are represented by potentials ρh_l and $\rho n_{l,nn}$ as given by formulae /37/ and /38/.

The corresponding Green function G acting on ρh_l and $\bar{G}$ on $\rho n_{l,nn}$ are as follows:

$$\left.\begin{aligned}G_\eta &= \frac{1}{2\pi}\frac{\bar{k}_o^2}{\rho\omega^2}\frac{\exp/i\bar{k}_oR_o/}{R_o},\\\bar{G}_\eta &= -\frac{\nu}{\pi}\frac{1}{\bar{\mu}S+2\nu^2}\frac{\exp/i\bar{k}_oR_o/}{R_o}\end{aligned}\right\}\quad /53/$$

$$\left.\begin{aligned}G_\psi &= \frac{\nu}{2\pi}\frac{1}{\bar{\mu}S+2\nu^2}\frac{\exp/i\bar{k}_oR_o/}{R_o},\\G &= -\frac{\rho\omega^2-\bar{k}_o^2}{2\pi k_o^2}\frac{1}{\bar{\mu}S+2\nu^2}\frac{\exp/i\bar{k}_oR_o/}{R_o}\\&+\frac{1}{2\pi S\,R_o}.\end{aligned}\right\}\quad /54/$$

where the integration contour related to the last term is the same as in Fig. 13.

At last the formal solutions for η_1 and ψ_1 are

$$\left.\begin{aligned}\eta_1 &= \frac{1}{2\pi\rho\omega^2}\bar{f}_p\, e_{prl}\left(\frac{\partial}{\partial x_r}\frac{1}{R_o}\right)\cdot\\ &\left[-i\bar{k}_oR_o\exp/i\bar{k}_oR_o/ + \exp/i\bar{k}_oR_o/ - 1\right]\\ &+\frac{\nu}{\pi\rho\omega^2 S}\bar{L}_{pk}\, e_{prl}\left(\frac{\partial^2}{\partial x_r\partial x_k}\frac{1}{R_o}\right)\cdot\\ &\left[-i\bar{k}_oR_o\exp/i\bar{k}_oR_o/ + \exp/i\bar{k}_oR_o/ - 1\right],\end{aligned}\right\} \quad /55/$$

$$\left.\begin{aligned}\psi_1 &= \frac{\nu}{2\pi\rho\omega^2 S}\bar{f}_p\, e_{prl}\left(\frac{\partial}{\partial x_r}\frac{1}{R_o}\right)\cdot\\ &\left[-i\bar{k}_oR_o\exp/i\bar{k}_oR_o/ + \exp/i\bar{k}_oR_o/ - 1\right]\\ &+\frac{/\rho\omega^2 - \mu\bar{k}_o^2/}{\pi\rho\omega^2 S}\bar{L}_{pk}\, e_{prl}\left(\frac{\partial^2}{\partial x_r\partial x_k}\frac{1}{R_o}\right)\cdot\\ &\left[-i\bar{k}_oR_o\exp/i\bar{k}_oR_o/ + \exp/i\bar{k}_oR_o/ - 1\right]\\ &-\frac{1}{4\pi S}\bar{L}_{pk}\, e_{prl}\left(\frac{\partial^2}{\partial x_r\partial x_k}\frac{1}{R_o}\right)R_o^2 .\end{aligned}\right\} \quad /56/$$

As we can see there are domains of the frequencies for which we observe the wave propagation $/k_o^2 > 0,\ \bar{k}_o^2 > 0/$, or exponential decrease $/k_o^2 < 0,\ \bar{k}_o^2 < 0/$. The formal solutions given above require still more detailed discussion on the frequency domains for which the waves propagate from a source.

11. TWO SOURCE MODELS

The classical source model of seismic event is represented by two dipoles of forces with opposite moments, so that the resulting rotation moment at the source vanishes $F_{i,k} + F_{k,i}$. In our symmetric micromorphic theory the corresponding model associated with internal structure can be also given by the symmetric body couple L_{ik}. It is an open question whether the source model given by body couple is accompanying the classical one or whether these models can be treated separately. In any case an important problem arises concerning the displacement fields $\underset{\sim}{u}$ related to both models. We suppose that these fields shall be the same and that the dipole of forces and the couple model differ only in the microdisplacement fields $\underset{\sim}{\phi}$.

The field $\underset{\sim}{u}$ is given by potentials ξ, $\underset{\sim}{\eta}$ and hence it is sufficient to compare these potentials for both models.

From eq. /49/ immediately follows that the parts related to the force and couple are in the constant ratio $S/2\nu$. The same ratio is easy to find in the eq. /55/. Thus, the $\underset{\sim}{u}$ fields related to both models are the same, apart from the factor $S/2\nu$. The microdisplacement fields evidently differ for both models.

The microdisplacement potentials φ and $\underset{\sim}{\psi}$ - eqs. /50/ and /56/ - contain also the terms responsable for standing waves. These waves are connected only with the part related to body couples $\underset{\sim}{L}$ acting in a source.

From the other point of view we shall remind that internal energy release is connected with high regional stress concentration and local weaknesses of structure. Thus, it seems reasonable to assume that body forces determined by the regional field are concentrated in a source. Assuming this point of view we shall put $\underset{\sim}{L} = 0$ evading at the same time the field of standing waves. We shall add here that another type of standing waves for some special values ω_o for which $k_o = 0$ are connected with internal structure, what is here connected with a certain natural length unit.

12. DISPLACEMENT AND MICRODISPLACEMENT FIELDS

The solutions /48, 49, 50/ and /55, 56/ enable to calculate displacement and microdisplacement fields according to relations /30/ and /31/. We have put $\underset{\sim}{L} = 0$ and the corresponding expressions for potentials depend only on body forces. Still the potentials contain factor:

$$-ik_oR_o \exp/ik_oR_o/ + \exp/ik_oR_o/ - 1$$

with one term constant. This term could give rise again for standing wave unless it will be compensated in the expressions for the $\underset{\sim}{u}$ and ϕ fields. In reality it is easy to proove this fact for displacement field, the constant terms from the ξ and $\underset{\sim}{\eta}$ potentials mutualy annihilate. For the microdisplacement field $\underset{\sim}{\phi}$ the same consideration lead us to condition $\varkappa = 0$ and hence to ascertain this condition we shall assume that material constants η and $\bar{\eta}$ vanish. Thus we have the condition, which has been already used when discussing plane wave problem:

$$\eta = \bar{\eta} = \tau = 0 , \quad \bar{\lambda} = \lambda$$

From this condition follows several consequences - compare eq. /40/:

$$\left. \begin{array}{l} K = N = 0 , \quad S = T = P , \quad H = 2\nu S , \\ M = /\lambda + 2\bar{\mu}/S + 4\nu^2 . \end{array} \right\} /57/$$

Further on we shall look on potentials φ and $\underset{\sim}{\Psi}$. It is easy to find now that - regarding - the microdisplacement field $\underset{\sim}{\phi}$ - the constant terms, which appeared in these potentials, vanish.

At last we can write the final expressions for potentials and the corresponding formulae for the displacements and microdisplacements.

The displacement potentials:

$$\xi = \frac{1}{2\pi\varrho\omega^2} \bar{f}_1 \left(\frac{\partial}{\partial x_1} \frac{1}{R_o} \right) \cdot \left[-ik_oR_o \exp/ik_oR_o/ + \exp/ik_oR_o/ - 1 \right] , \quad /58/$$

$$\eta_1 = \frac{1}{2\pi\varrho\omega^2} \bar{f}_p \, e_{prl} \left(\frac{\partial}{\partial x_r} \frac{1}{R_o} \right) \cdot \left[-i\bar{k}_oR_o \exp/i\bar{k}_oR_o/ + \exp/i\bar{k}_oR_o/ - 1 \right] . \quad /59/$$

The displacement field:

$$u_k = \frac{1}{2\pi\varrho\omega^2} \left[\bar{f}_1 \left(\frac{\partial^2}{\partial x_1 \partial x_k} \frac{1}{R_o} \right) + \bar{f}_1 \left(\frac{\partial}{\partial x_1} \frac{1}{R_o} \right) \frac{\partial}{\partial x_k} \right] \cdot \left[-ik_oR_o \exp/ik_oR_o/ + \exp/ik_oR_o/ + i\bar{k}_oR_o \exp/i\bar{k}_oR_o/ - \exp/i\bar{k}_oR_o/ \right] . \quad /60/$$

The microdisplacement potentials:

$$\mathcal{H} = 0 , \quad /61/$$

$$\varphi = \frac{\nu}{\pi\varrho\omega^2 S} \bar{f}_1 \left(\frac{\partial}{\partial x_1} \frac{1}{R_o} \right) \cdot \left[-ik_oR_o \exp/ik_oR_o/ + \exp/ik_oR_o/ - 1 \right] , \quad /62/$$

$$\Psi_1 = \frac{\nu}{2\pi\varrho\omega^2 S} \bar{f}_1 \, e_{pr1} \left(\frac{\partial}{\partial x_r} \frac{1}{R_o} \right) \cdot \left[-i\bar{k}_o R_o \exp/i\bar{k}_o R_o/ + \exp/i\bar{k}_o R_o/ - 1 \right] . \qquad /63/$$

The microdisplacement field:

$$\phi_{ab} = \frac{\nu}{\pi\varrho\omega^2 S} \left[\bar{f}_1 \left(\frac{\partial^3}{\partial x_1 \partial x_a \partial x_b} \frac{1}{R_o} \right) \cdot + \bar{f}_1 \left(\frac{\partial^2}{\partial x_1 \partial x_a} \frac{1}{R_o} \right) \frac{\partial}{\partial x_b} + \bar{f}_1 \left(\frac{\partial^2}{\partial x_1 \partial x_b} \frac{1}{R_o} \right) \frac{\partial}{\partial x_a} + \bar{f}_1 \left(\frac{\partial}{\partial x_1} \frac{1}{R_o} \right) \frac{\partial^2}{\partial x_a \partial x_b} \left[-ik_o \exp/ik_o R_o/ + \exp/ik_o R_o/ + ik_o \exp/ik_o R_o/ - \exp/ik_o R_o/ \right] \right. \qquad /64/$$

where: $k_o^2 = \dfrac{\varrho\omega^2 S}{/\lambda + 2\mu/S + 4\nu^2}$, $\bar{k}_o^2 = \dfrac{\varrho\omega^2 S}{\bar{\mu}S + 2\nu^2}$,

$$S = \varrho\omega^2 I - 2\bar{\nu}$$

For geophysical applications we very often take only the 1/R approximation omitting higher orders terms. For the displacement and microdisplacement field we would get:

$$u_k = -\frac{1}{2\pi\varrho\omega^2} \bar{f}_1 \frac{x_k x_1}{R_o^3} \left[k_o^2 \exp/ik_o R_o/ - \bar{k}_o^2 \exp/i\bar{k}_o R_o/ \right] \exp/-i\omega t/ \quad , \qquad /65/$$

$$\phi_{ab} = \frac{\nu}{\pi\varrho\omega^2 S} \bar{f}_1 \frac{x_1 x_a x_b}{R_o^4} \left[k_o^3 \exp/ik_o R_o/ - \bar{k}_o^3 \exp/i\bar{k}_o R_o/ \right] \exp/-i\omega t/ \quad . \qquad /66/$$

REFERENCES

1. Ben-Abraham S.I., 1970, Generalized Stress and Non-Riemannian Geometry, Fundamental Aspects of Dislocation Theory, /Nat. Bur. Stand. /U.S./, Spec. Publ. 317, II, 1970/, 943 - 962.
2. Davison Ch., 1927, The Founders of Seismology, Cambridge, University Press.
3. Eringen A.C., 1968, Theory of Micropolar Elasticity, Fracture, ed. H.Liebowitz, Acad. Press, New York and London, Vol. 2 /1968/, 621 - 729.
4. Eringen A.C., Claus W.D., Jr., 1970, A Micromorphic Approach to Dislocation Theory and its Relation to Several Existing Theories, Fundamental Aspects of Dislocation Theory, /Nat. Bur. Stand. /U.S./ Spec. Publ., 317, II, 1970/, 1023 - 1040.
5. Eringen A.C., Suhubi E.S., 1964, Nonlinear Theory of Simple Micro-Elastic Solids - I, Int. J. Engng. Sci. Vol. 2, Pergamon Press, 189 - 203.
6. Gutenberg B., 1926, Grundlagen der Erdbebenkunde, Univ. Frankfurt a/M.
7. Hanyga A., Teisseyre R., 1973, The Fundamental Source Solutions in the Symmetric Micromorphic Continuum, Rivista Italiana Di Geofisica, Vol. 23, No. 5/6, 336 - 340.
8. Hanyga A., Teisseyre R., 1974, Point Source Models in the Micromorphic Continuum, Acta Geophys. Pol., Vol. 22, No. 1, 11 - 20.
9. Hobbs W.H., 1907, Earthquakes, New York, Appleton and Co.
10. Hordejuk J., Droste Z., Teisseyre R., 1974, An Analysis of Rapid Changes in Earth Tilts and Displacements Accompanying Seismic Shocks Recorded by Mechanical and Galvanometric Seismographs, Publ. Inst. Geoph., Pol. Ac. Sci. /in print/.
11. Imamura A., 1937, Theoretical and Applied Seismology, Tokyo, Maruzen Co.
12. Kröner E., Seeger A., 1959, Nicht-lineare Elastizitätstheorie der Versetzungen und Eigenspannungen, Arch. Rational Mech. Anal. 3, 97 - 119.
13. Suhubi E.S., Eringen A.C., 1964, Nonlinear Theory of Micro-Elastic Solids - II, Int. J. Engng. Sci. Vol. 2, Pergamon Press, 389 - 404.
14. Teisseyre R., 1969, Dislocational Representation of Thermal Stresses, Acta Geophys. Pol. 17, 3 - 12.
15. Teisseyre R., 1973, Earthquake Processes in a Micromorphic Continuum, Pure Appl. Geophys. Vol. 102 /1/, 15 - 28.

SURFACE DEFORMATION IN ICELAND AND CRUSTAL STRESS OVER A MANTLE PLUME

Eysteinn Tryggvason

Department of Earth Sciences, The University of Tulsa,
Tulsa, Oklahoma, U.S.A.

ABSTRACT. The horizontal radial flow of plastic material near the top of a mantle plume excerts stresses on the lower lithosphere. The velocity of this flow is a function of the distance from the center of the plume. The radial stress in the lithosphere will be tensional over the plume but compressional outside of the plume, while the tangential stress is tensional throughout the whole region of the radial flow. The difference between radial and tangential stress represents a shear stress. This has a maximum a short distance outside of the plume. If both radial and tangential stress is tensional, as above the plume, the surface deformation will be related to normal faults. High volcanic activity is expected but seismic activity will be low and only small earthquakes will occur because of the low tensile strength of crustal rocks. Where the radial stress is compressional and the tangential stress is tensional, as in areas outside of the plume, the ground deformation will be largely associated with strike-slip faults. Volcanic activity is expected to be low, but seismic activity may be high with occasional large earthquakes because of the relatively great shear strength of crustal rocks.

INTRODUCTION

The driving mechanism for the plate tectonics is generally assumed to be convection currents in the earth's mantle. The rising currents are supposedly concentrated in a number of mantle plumes or "hot spots" beneth or near the world's rift zones (Morgan, 1971). These mantle plumes are expressed on the earth's surface by recent volcanic activity,

Thoft-Christensen (ed.), Continuum Mechanics Aspects of Geodynamics and Rock Fracture Mechanics, 245-254.

not associated with andesitic trench type activity, and by surface elevation, higher than the surrounding areas (Morgan, 1972).

The total number of mantle plumes was estimated as about 20 by Morgan (1972), but there may be some plumes which have not been positively identified because of questionable surface expressions. One of the most convincing surface expression of a mantle plume is in Iceland. The high volcanic activity associated with roughly circular topographic high, located on the Mid-Atlantic Ridge, all support the hypothesis of a mantle plume centered below Iceland.

THE MANTLE FLOW

The mantle plume contains mantle material of relatively low viscosity and density, flowing upward at velocity which may be about 2 meter per year (Morgan, 1972). At the top of the plume the flow becomes horizontal and outward from the center of the plume forming the asthenosphere, a layer of low viscosity near the top of the mantle (Figure 1). The boundary of the plume and of the asthenosphere is expected to be gradual with lowest viscosity in the center of the plume.

The velocity of the radial flow in the asthenosphere must be zero above the center of a radially symmetrical plume, increasing outward to a maxima at a radial distance similar to the radius of the plume. At great distances from the plume the velocity of the radial flow decreases asymptotically to zero. The unknown viscosity and density variations within the plume and the asthenosphere and probable variations in the plume radius with depth and in asthenosphere thickness with the distance from the plume make it impossible to construct any accurate and reliable flow model. However, we will assume that the horizontal component of the mantle flow in the asthenosphere is purely radial with zero velocity above the center of the plume and inversely proportional to the radial distance at great distances from the plume and with one maxima at an intermediate distance.

In search of a mathematical expression fulfilling the above criteria, two simple expression were adopted as reasonable approximations of the actual velocity V of the radial flow:

a. $$V = 2V_o X/(1 + X^2) \tag{1}$$

and

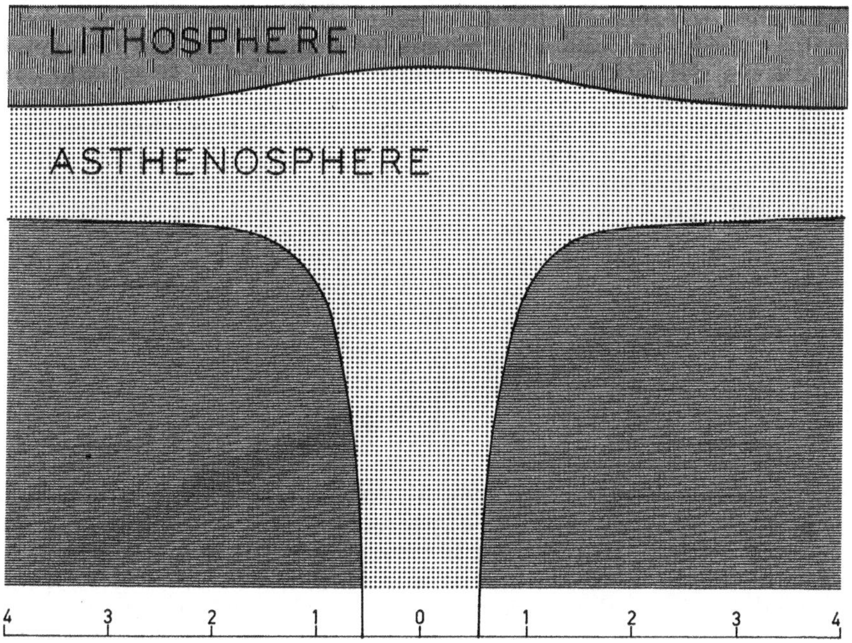

Figure 1. A schematic model of the top of a mantle plume showing the low viscosity plume material by dot-shading and the normal mantle material by horizontal line shading. Relative thickness of the asthenosphere is based on the flow model of Figure 3. Horizontal scale is in units of X (see text).

b. $$V = 2V_o X^2/(1 - X + X^2 + X^3) \quad (2)$$

where $X = r/R$, r is the radial distance from the center of the plume and R is the distance to the point of maximum radial velocity. V_o is the maximum radial velocity.

HORIZONTAL STRESS IN THE EARTH'S CRUST ABOVE A MANTLE PLUME

The horizontal flow of a uniform viscous asthenosphere will exert a force on the overlying lithosphere, proportional to the velocity of the asthenosphere relative to the lithosphere and in the direction of the relative movement. If the thickness of the lithosphere is small compared to the horizontal dimensions of the flow regime, the variations of the horizontal components of stress with depth in the lithosphere will be small. Under that condition the tensile stress in a stationary lithosphere is given by:

$$S_r = a \frac{dV}{dX} \qquad (3)$$

and

$$S_t = a \frac{V}{X} \qquad (4)$$

where S_r and S_t are the radial and tangential components of horizontal tensile stress respectively and a is a proportionality constant which depends on the viscosity and thickness of the asthenosphere. The total horizontal tensional stress S_{tot} is defined here as the sum of the two principal horizontal stress components ($S_{tot} = S_r + S_t$) and the horizontal shear stress S_{sh} as the difference between these stress components ($S_{sh} = S_t - S_r$).

Figures 2 and 3 show the variation of the radial velocity and of the various stress components with distance from the center of the plume for the two velocity functions (1), and (2). One of the velocity functions (Figure 2) gives maximum tensional stress above the center of the plume while the other function (Figure 3) gives zero tensional stress at the same location.

Neither of these condition is likely to exist in nature. The true stress condition in a stationary lithosphere above a mantle plume may be somewhere between those shown on figures 2 and 3, with the principal characteristics as follows:

Radial stress is tensional at distances less than R_o and compressional at greater distances. Maximum radial tension is expected at a radial distance of about 0.3 R_o and maximum radial compression at about 1.5 R_o distance.

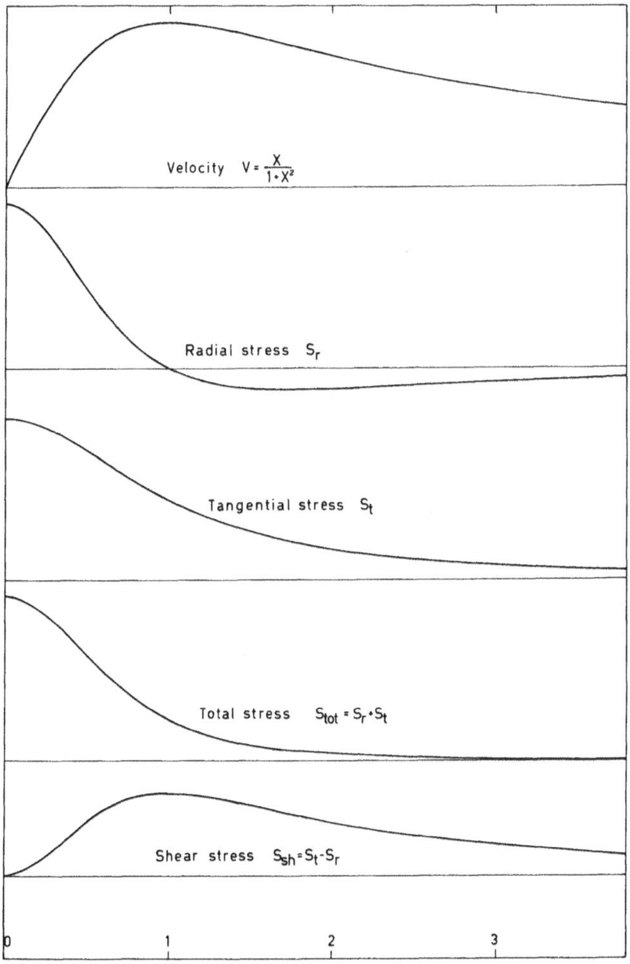

Figure 2. Horizontal asthenosphere velocity and components of tensile stress in the lithosphere above a mantle plume, based on velocity function (1). Vertical scales are arbitrary but the horizontal scale gives the distance from the center of the plume in units of X (X = r/R where R is the radial distance to the point of maximum horizontal velocity in the asthenosphere).

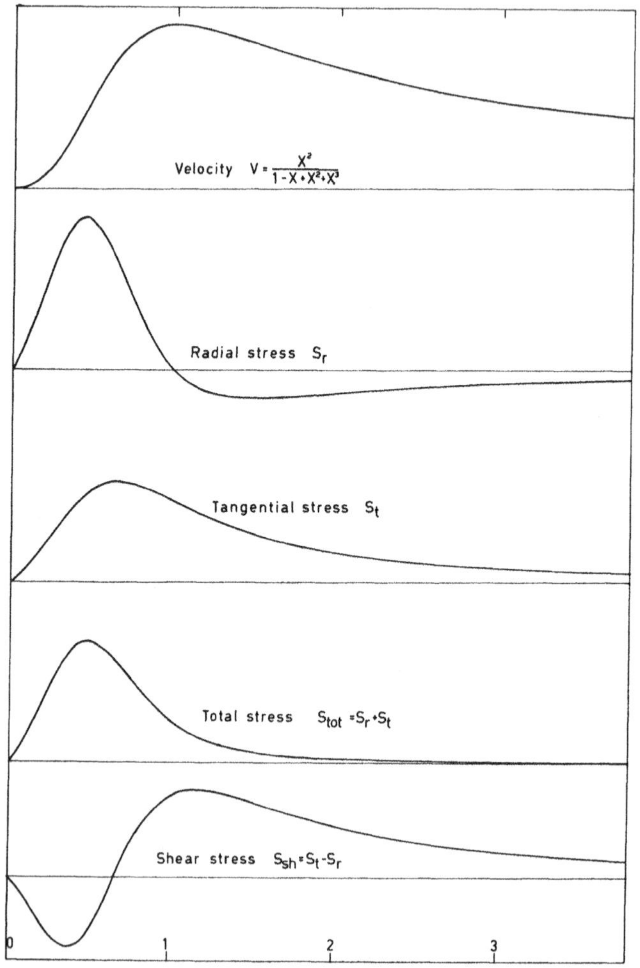

Figure 3. Horizontal asthenosphere velocity and components of tensile stress in the lithosphere above a mantle plume, based on velocity function (2).

Tangential stress is tensional over the whole flow regime with a maxima at radial distance of about 0.5 R_o.

Total tension exists over the whole flow regime of the plume with a maxima at about 0.4 R_o radial distance from the center of the plume and near zero value at distances greater than 1.5 R_o.

Shear stress is zero over the center of the plume and probably very small at radial distances less than 0.5 R_o. Maximum shear stress is expected at about 1.1 R_o radial distance and it decreases slowly at greater distances.

SURFACE EXPRESSION OF THE STRESS FIELD

When the earth's crust is exposed to continous stress over a long period of time, it will gradually be deformed. The deformation that takes place will depend on the nature of the stress and on the physical characteristics of the crustal material (de Sitter, 1964).

If both principal components of the horizontal stress are tensional, normal faulting is likely to occur with the strike of the faults nearly perpendicular to the direction of maximum tension. Open fissures and grabens will form and conditions for volcanic activity are favorable. The deformation will take place without significant earthquake activity because of the low tensile strength of the crustal rocks.

If one principal component of the horizontal stress is tensional and the other is compressional, shear faults are favored with the strike of the faults at an angle of 30° to 45° to the direction of maximum compression. Vertical component of the deformation should be small and open fissures are unlikely. An echelon fault pattern is likely, especially if the total stress is tensional. The deformation is likely to be associated with occational large earthquakes, because of the relatively high shear strength of the crustal rock. Conditions for volcanic activity are less favorable than in areas of pure tension.

If both principal components of horizontal stress are compressional, folding and reverse faults are likely to occur. This condition is not expected in the vicinity of a mantle plume and, therefore, not discussed further here.

The principal features of the surface deformation that

will occur in stationary lithosphere above a mantle plume can be predicted from the assumed stress field.

Close to the center of the plume, no significant deformation is expected, although normal faults may occur.

At about 0.3 R_o radial distance we may expect normal faults with strikes in tangential direction, but this is rather uncertain.

At radial distances between 0.4 R_o and 0.9 R_o the deformation will be dominated by normal faults and grabens with strikes in approximately radial direction. An echelon fault pattern is likely near the outer boundary of this zone.

At radial distances between 1.1 R_o and 2 R_o the surface deformation will be dominated by shear faults with strikes at an angle of 30° to 45° to the radial direction. The faulting will be most intense near the inner boundary of this zone.

Volcanic activity will be higher at radial distances less than R_o than at greater distances.

Earthquake activity will be insignificant at radial distances less than R_o. Large earthquakes are most likely at radial distances slightly greater than R_o.

ICELAND IN THE MANTLE PLUME MODEL

If the mantle plume model described above is approximately correct the location of the center of the Iceland mantle plume can be determined, and also the distance R_o from this center to the circle of maximum radial velocity in the asthenosphere.

Figure 4 shows the principal seismic and volcanic zones of Iceland. A circle is drawn in such a way that as much as possible of the volcanic zones fall inside the circle while all areas where earthquakes of magnitude greater than 6 are known to occur (Tryggvason, 1973) fall outside this circle. This circle has a radius of 140 kilometers, which is assumed to be equal to R_o in our model. The center of this circle is at about 64°45'N, 18°00'W or very close to the geographical center of Iceland. The Iceland mantle plume is assumed to be centered at this location.

Only a portion of the tectonic phenomena in Iceland can be explained by our model of stationary lithosphere over a mantle plume. The fact that Iceland is located on the boundary of two moving lithosphere plates must cause a deviation from circular symmetry of the crustal stress field.

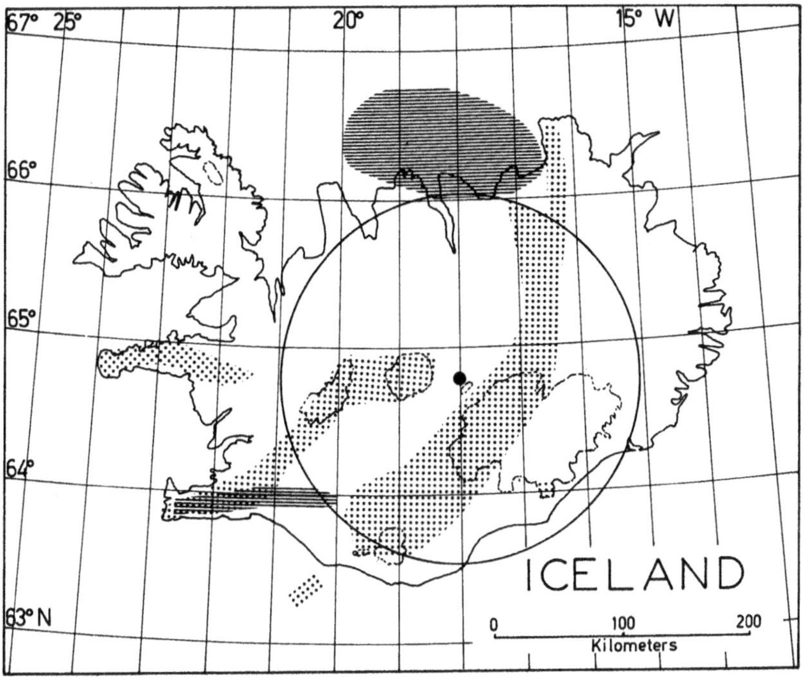

Figure 4. Map of Iceland showing zones of recent volcanic activity (dotted) and zones where earthquakes of magnitude exceeding 6 occur (horizontal shading). The dot near 64°45'N, 18°00'W is the assumed center of the Iceland mantle plume and a circle of 140 kilometer radius shows where maximum radial velocity of asthenosphere flow is predicted.

REFERENCES

de Sitter, L. U. Structural geology, 2nd ed., 551 pp, McGraw-Hill, New York, 1964.

Morgan, W. J. Convection plumes in the lower mantle, Nature, 230, pp 42-43, 1971.

Morgan, W. J. Deep mantle concection plumes and plate motion, Am. Assoc. Petroleum Geologists Bull., 56, pp 203-213, 1972.

Tryggvason, Eysteinn. Seismicity, earthquake swarms, and plate boundaries in the Iceland region, Bull. Seismol. Soc. Amer., 63, pp 1327-1348, 1973.

FAULT DISPLACEMENT AND GROUND TILT DURING SMALL EARTHQUAKES*

Eysteinn Tryggvason

Department of Earth Sciences, University of Tulsa, Tulsa, Oklahoma, U.S.A.

ABSTRACT. A program of repeated precision leveling at selected locations in Iceland over a period of seven years shows an annual rate of tilting of less than one microradian per year on the flanks of the active rift zones. Earthquakes of magnitude 5 to 6 are associated with measurable fault displacement at distances up to 25 kilometers from the epicenters and the area around the active fault is abnormally deformed during the earthquakes. During a period of one or two years after an earthquake the rate of ground deformation is higher than normal. This is explained by plastic flow at depth which tends to restore the surface to the position before the sudden deformation.

INTRODUCTION

Significant ground deformation has frequently been observed to take place during large earthquakes. This deformation is usually associated with surface faulting and the two sides of the faults have moved relative to each other. A detailed study of the deformation which took place during the California earthquake of April 18, 1906 (Lawson et al., 1908) resulted in the elastic rebound theory of earthquakes (Reid, 1910) which assumes that tectonic earthquakes are the result of elastic strain, greater

* This work has been supported by grants GP-5365, GA-987, GA-4112, GA-24152 from the National Science Foundation, Washington.

Thoft-Christensen (ed.), Continuum Mechanics Aspects of Geodynamics and Rock Fracture Mechanics, 255-269.

than the strength of the rock can withstand, producing sudden relative displacements of neighboring portions of the earth's crust. This theory is supported by numerous studies of ground deformation during large earthquakes, and as there is no indication of significant difference between the focal mechanism of great and small earthquakes, except the magnitude of the relative movement, small earthquakes are also assumed to be caused by sudden relative displacement of the earth's crust. This displacement is normally associated with faults and most focal mechanism models assume that one side of the fault is permanently displaced relative to the other side during an earthquake.

The area of the fault plane where the dislocation takes place during an earthquake has been the subject of numerous studies, and also the amount of displacement, but the relation between these parameters of earthquakes and their magnitudes has not been firmly established (Wyss and Brune, 1968; and Trifunac, 1974). However, both the area of the fault plane and the displacement normally increase with increased earthquake magnitude. The dislocation which supposedly takes place on a fault plane during an earthquake must vary from place to place, with one spot of maximum dislocation and diminishing dislocation towards the edge of the displaced fault plane. If the dislocation is known and the earth material behaves as uniform, homogeneous elastic solid, the displacement field can be computed (Haskell, 1969). However, the earth is not homogeneous and the distribution of dislocation in the plane of an earthquake fault is normally not known, so theoretical computations of the deformation must be subject to errors.

Although most models of earthquake foci assume that the relative movement of the two sides of a fault during an earthquake is parallel to the fault plane, this may not always be true. The two side may be moving away from each other. This type of movement will result in open fissure, but such fissure may possibly be closed immediately by elastic deformation due to the normal compressional stress caused by the weight of the overburden. The end result will then be a zone of abnormally low compressional stress in the vicinity of the fault plane. Similarily it may be possible that the two sides of the fault move towards each other during an earthquake, creating a zone of abnormally high compressive stress. Such zones of abnormal compressional stress will exert forces on the surrounding rock. These forces may cause gradual deformation over an extended period of time after an earthquake, if the rock yields to forces of long duration. This aftereffect of earthquakes, if it exists, is partly controlled by the anelastic properties of the rocks, such as plasticity or viscosity.

There are very few recorded observations of ground deformation associated with earthquakes in Iceland, although fissures in the ground have been observed on a number of occasions (Thoroddsen, 1925).

During an earthquake in 1789, the Thingvellir graben subsided about 60 centimeters, judged from observations made about three years after the earthquake on the principal faults on both sides of the graben (Thoroddsen, 1925). The earthquakes in South Iceland in 1896 and 1912 left fissures in the ground where some right lateral strike-slip displacement had taken place, but the amount of this displacement cannot be accurately determined (Thoroddsen, 1925; Einarsson, 1967; Tryggvason, 1967).

These few observations of fault displacement in Iceland during recent earthquakes indicate that considerable deformation has taken place in past earthquakes.

OBSERVATIONS

When the present program of repeated precision leveling was started in 1966, the primary aim was to determine the rate of vertical ground deformation within the zone of present tectonic activity, which is assumed to coincide with the zone of recent volcanism. The following criteria were adopted for the selection of locations for the leveling profiles:

a. They should lie within or near the zone of recent volcanic activity.
b. They should be in an area of recent crustal deformation.
c. The ground surface should allow easy construction of stable bench marks.

With these requirements in mind, several locations on the Reykjanes Peninsula were visited and the old postglacial lava field southeast of the small community Vogar was the first location selected, after the Reykjanes area had been rejected due to adverse weather conditions. In light of what happened later at Reykjanes, the earthquake swarm of September 1967 and associated changes in the Reykjanes geothermal field (Tryggvason, 1970), it is certainly regrettable that no precision leveling was made there. The second location selected for repeated precision leveling was the Búrfellshraun lava about 6 kilometers southwest of the town of Hafnarfjördur (Figure 1). More locations for precision leveling were added later, but those are not considered in this paper.

The standard error of leveling has been approximately $0.2\sqrt{D}$ millimeters where D is the distance along the profile of leveling in kilometers (Tryggvason, 1968). The ground deformation is determined from the comparison of two levelings, usually made one or more years apart. The error in the observed deformation is then caused by the measuring errors of the two levelings and also by any movements of the bench marks relative to the near surface formations which deformation is to be determined.

Comparison of two levelings made one or a few days apart with two levelings made one or more years apart on the same profile gives the magnitude of irregular movements of the individual bench marks. The standard deviation of single bench mark elevation from estimated elevation, using surrounding bench marks, is 0.10 to 0.15 millimeters, when the two levelings are made one year apart. Levelings made two or more years apart show no obvious increase in this irregular bench mark movement.

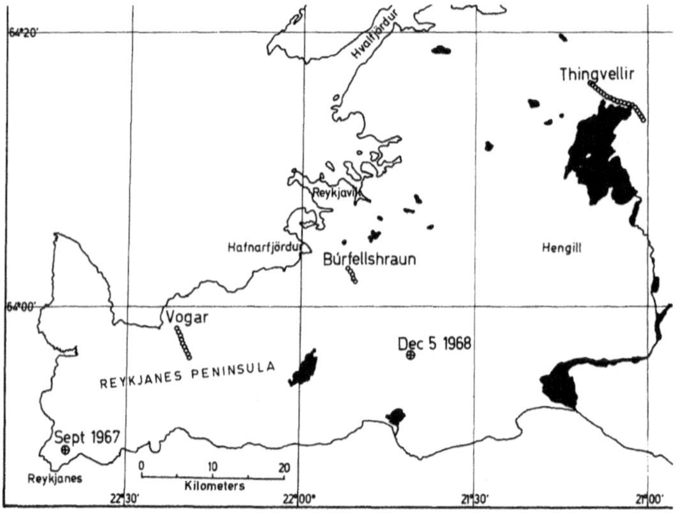

Figure 1. Location of the precision leveling profiles Vogar and Búrfellshraun in Southwest Iceland and approximate location of the earthquake swarm of September, 1967 and the earthquake of December 5, 1968.

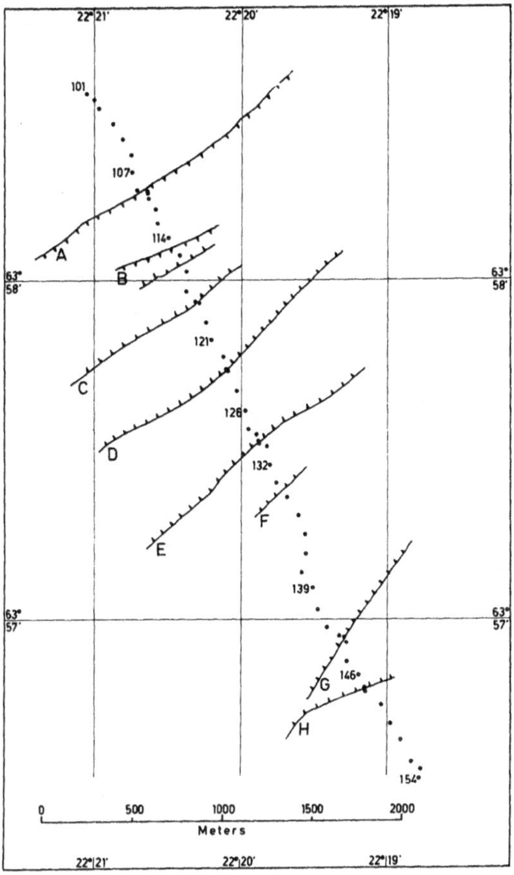

Figure 2. The Vogar precision leveling profile on the Reykjanes Peninsula in Southwest Iceland. Bench mark number is given for a few of the bench marks. Principal faults which cross the profile are shown, but numerous fissures with no observable vertical displacement are not shown.

The standard error of observed ground tilt over a period of one or more years, due to leveling errors and irregular bench mark movements is approximately $0.4\sqrt{D}$ millimeters. At the 95 percent confidence level the error in relative vertical displacement is then roughly $0.8\sqrt{D}$ millimeters and at the 99 per cent confidence level the error is about $1.0\sqrt{D}$ millimeters.

The two precision leveling profiles that will be discussed here are located on the Reykjanes Peninsula in Southwest Iceland (Figure 1). Both are on the northwest flank of the zone of recent faulting and volcanism, but this zone is a direct continuation of the seismically active Reykjanes Ridge. Each profile consists of a number of permanent bench marks along somewhat irregular line roughly perpendicular to the principal direction of faulting.

The Vogar profile (Figure 2) has 54 bench marks along a 4200-meter long line which crosses a narrow zone of intense faulting in early postglacial lava of the Strandarheidi shield volcano. The age of this lava has been estimated as 10,000 years, but this estimate is very inaccurate (Tryggvason, 1970). The southwest end of this zone of intense faulting is near Reykjanes, about 25 kilometers southwest of the profile, where a high temperature geothermal field is located (Pálmason, 1967), and where earthquake swarms occur very frequently (Tryggvason, 1973).

The Búrfellshraun profile (Figure 3) has 32 bench marks along an irregular line about 2500 meters long, located about 6 kilometers southeast of the town of Hafnarfjördur in Southwest Iceland. The volcano Búrfell lies about 0.5 kilometers east of the southeast end of the profile. The profile is located on the lava from Búrfell and the age of this lava has been determined as about 7000 years (Kjartansson, 1973).

The Búrfellshraun profile lies near the northeast end of a narrow zone of intense faulting and recent volcanic activity. The southwest end of this fault zone is near Krisuvik, about 25 kilometers southwest of the profile, where there is a high temperature geothermal field (Pálmason, 1967) and where earthquakes are more frequent than at any other location in Iceland (Tryggvason et al., 1958).

The Vogar profile was leveled in 1966, 1968, and 1971, and about 1/3 of the profile was also leveled in 1969. The Búrfellshraun profile was leveled once every year from 1966 to 1970, but the result of the first three levelings did not indicate significant ground deformation, and are therefore not considered here.

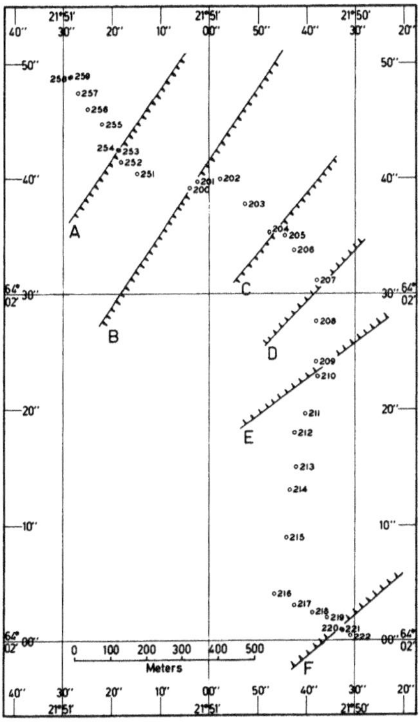

Figure 3. The Búrfellshraun precision leveling profile near Hafnarfjördur in Southwest Iceland. Principal faults crossing the profile are shown with direction of vertical displacement indicated by hachures.

RESULT OF REPEATED LEVELING

The Vogar profile

The first leveling on the Vogar profile was made during the month of June, 1966, and the exact elevation differences of the bench marks determined. The leveling showed clearly eight faults with vertical displacement of 2 to 11 meters and tilting of the blocks between the faults (Figure 4). Numerous fissures with no or very small vertical displacement were observed in the field. As the profile is on a gently sloping shield volcano Strandarheidi, the ground tilt before faulting can be estimated (Tryggvason, 1970). The total vertical ground displacement since the Strandarheidi eruption is considerably greater than the fault displacements indicate, because of the tilting of the blocks between the faults. The downthrown side of fault G has probably subsided 40 to 50 meters relative to the upthrown side of fault A (Tryggvason, 1970). If the estimated 10,000 year age of the Strandarheidi lava is correct and if the deformation has been going on at a constant rate during these 10,000 years, the average annual subsidence has been about 5 millimeters. A subsidence of this magnitude would be easily observed by repeating the precision leveling after one year.

The second leveling of the Vogar profile was made in June, 1968. Very significant vertical ground movement was observed over the two year period since the first leveling (Figure 4), but the Reykjanes earthquake swarm of September, 1967 had occurred and considerable vertical movement was seen in the epicentral area, about 25 kilometers to the southwest of the Vogar profile (Tryggvason, 1970). As the faults of visible vertical displacement in the epicentral region had strikes that pointed towards the Vogar profile, it was assumed that some fraction of the observed deformation on the Vogar profile had taken place during the 1967 Reykjanes earthquake swarm.

A study of the 1966-1968 deformation of the Vogar profile (Tryggvason, 1970) indicates that at least three processes were involved. A general tilt of the area of about 2 microradians down to the southeast, subsidence of the northwestern part of the profile centered near fault A (Figure 2), and dip-slip movement on the fault A. It may be assumed that the deformation which can be correlated with the fault A is directly caused by the Reykjanes earthquake swarm while the general tilting may be unrelated to the swarm. Leveling of a fraction of the profile in August, 1969 seems to support this hypothesis as no additional displacement was observed on fault A and the subsidence of the region around fault A had been reversed (Tryggvason, 1970).

The profile was leveled again in July, 1971. Comparisons of the 1968 and 1971 levelings (Figure 4) shows very little correlation with the changes observed between 1966 and 1968, except for the regional tilt towards southeast which continues. However the tilt from 1968 to 1971 is only about one microradian in three years while a tilt of about two microradians was observed in two years, 1966 to 1968. The zone around fault A which had subsided

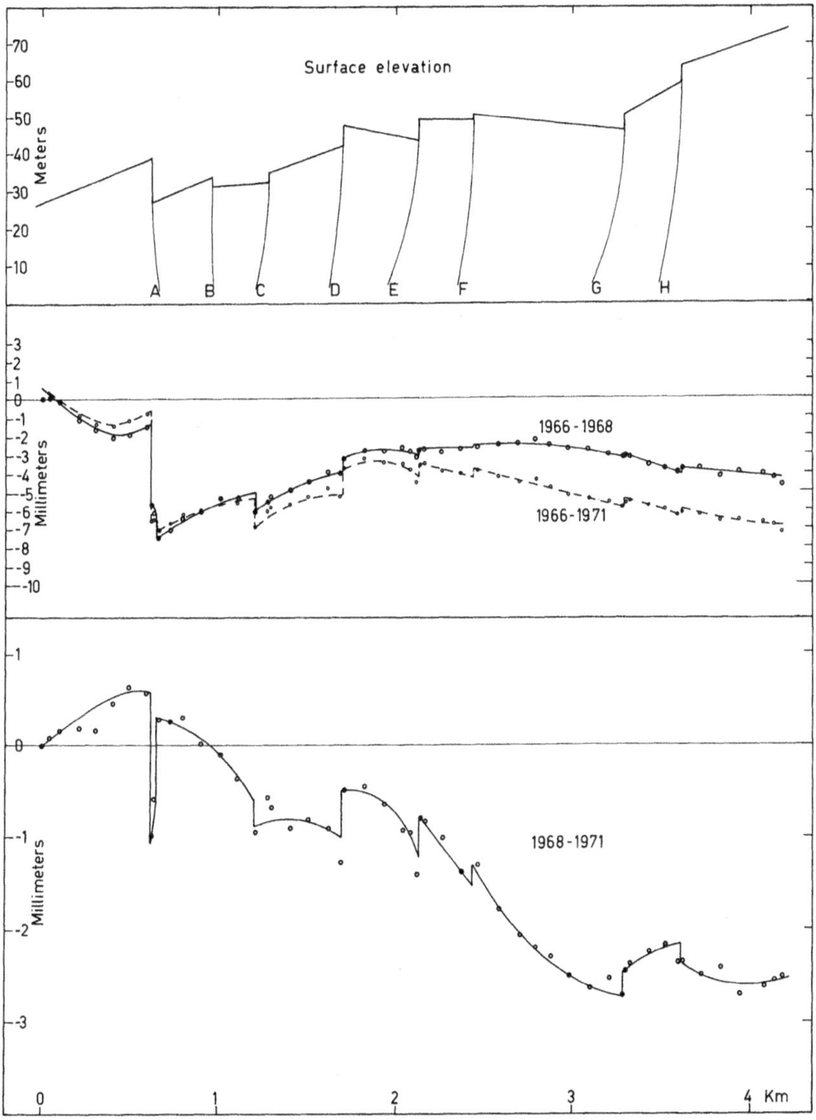

Figure 4. Result of repeated precision leveling of the Vogar profile. Northwest end of the profile is to the left (see Figure 2). The vertical displacement (central and lower section) is relative to the northwest end of the profile.

about 8.7 millimeters between 1966 and 1968 (Tryggvason, 1970) had risen almost one millimeter between 1968 and 1971. The part of the profile which lies to the southeast of fault A had been bent uniformly between 1966 and 1968 due to the subsidence and dip-slip movement on fault A. Between 1968 and 1971 this part of the profile was slightly bent in the opposite direction (concave on Figure 4), and minute dip-slip movements are indicated on most of the faults of the region (Figure 4). A tilted block with two bench marks (109 and 110) on the down thrown side of fault A is anomalous. From 1966 to 1968 this block subsided about 2 milli meters less than bench marks father to the southeast, but they continued to subside after 1968.

The Búrfellshraun profile

The Búrfell lava (Búrfellshraun) on which the profile lies is rather porous and by no means as competent as the Strandarheidi lava on which the Vogar profile was laid. The surface appearance of this lava indicates lesser age than many of the other lavas. The profile was laid along a very prominent lava channel from the volcano Búrfell. The shape of this lava channel determines the layout of the profile (Figure 3). At the time of the Búrfell eruption, about 7,000 years ago, the bottom of the lava channel must have had a uniform slope away from the volcano or to the left on Figure 5.

The Búrfellshraun profile was first measured in 1966 and again in 1967, 1968, 1969, and 1970. The leveling shows six definite faults with vertical displacement (Figure 5) and the subsidence of the central part of the profile relative to its northwest end is at least 10 meters and may be as much as 20 meters during the last 7,000 years. This indicates a subsidence rate of 1.5 to 3 millimeters per year if the subsidence is a continuous process.

No significant movement was observed during the first two years, 1966 to 1968. However, at least 4 of the 32 bench marks had been uplifted approximately one millimeter, obviously due to some surface processes unrelated to crustal deformation (Tryggvason, 1968). As the bench marks were not as stable as on the Vogar profile, the error in comparison of two levelings was greater. However, the maximum ground movement on the profile between 1966 and 1968 can not have exceeded 0.5 millimeters, or the average rate of vertical movement of one part of the profile relative to

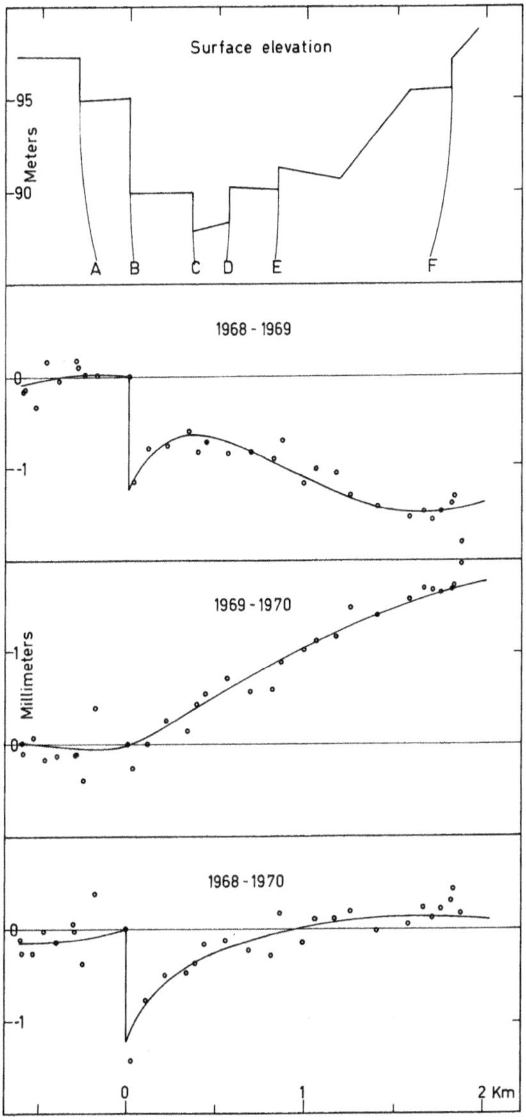

Figure 5. Result of repeated precision leveling of the Búrfell-shraun profile. Northwest end of the profile is to the left. The vertical movement is relative to bench mark 200 on the up thrown side of fault B (see Figure 3).

any other part of the profile must have been less than 0.25 millimeters per year, or only a small fraction of the average subsidence over the last 7,000 years.

The leveling in July, 1969 showed that a dip-slip displacement had taken place on fault B during the preceeding year and that the region to the southeast of this fault had been significantly deformed (Figure 5). The leveling in June, 1970 showed no additional displacement on fault B, but a reversal of the deformation to the southeast of this fault had taken place. The total observed deformation between 1968 and 1970 is thus a dip-slip displacement of about 1.3 millimeters on the fault B and a bending of a 500 meters wide zone to the southeast of this fault, but no significant deformation farther to the southeast, nor to the northwest of fault B. There is no indication of displacement on any of the faults crossing the Burfellshraun profile, except fault B.

DISCUSSION

The surface deformation that took place in the vicinity of fault A on the Vogar profile during the Reykjanes earthquake swarm of September, 1967 was partly a dip-slip movement on the fault and partly a subsidence of the fault zone. The mechanism which can produce this type of deformation has been described as tensional failure at depth and dip-slip failure near the surface (Tryggvason, 1970). The tensional failure at depth may have separated the two sides of the fault during the earthquake, but the displacement of one side of the fault relative to the other side was so small that elastic deformation, due to the pressure of the overburden, closed the fault again, except in the surface layer where the overburden pressure is small. The effect of this type of tensional failure will be a zone at depth where the compressional stress is abnormally low, lower than the hydrostatic pressure of the overburden, and a dip-slip displacement near the surface.

The anomalous deformation that was observed in the vicinity of fault A after the earthquake continued for a period of one or two years. This is here interpreted as a delayed effect of the earthquake, caused by slow plastic movement or creep which tends to eliminate the stress anomalies which were caused by the earthquake. The small uplift of the area around fault A between 1968 and 1971 (Figure 4) may be due to plastic flow in the lower part of the crust towards the zone of abnormally low pressure. Leveling of a part of the profile in 1969 (Tryggvason, 1970) shows that most of the uplift around fault A between 1968 and 1971 did take place during the first year of this three year period, so the relaxation time for this process is of the order of one year. This means that the zone around fault A may have been uplifted several millimeters between the time of the earthquake swarm in September, 1967 and the time of the second leveling in June, 1968.

The very small displacements on faults D, E, F, G, and H, and possibly on one additional fault, between the levelings of 1968 and 1971 (Figure 4) may be interpreted as the result of tension due to the plastic flow at depth towards fault A. This has caused small (about one millimeter or less) tensional displacements in the surface layer, but uniform tensional stress at depth. The problem of how deep the fault displacement extends into the ground has not been solved. A model study of the displacement of fault A indicated that dip-slip displacement on that fault extended to a depth of approximately 300 meters (Tryggvason, 1970). The much smaller dip-slip displacement on the other faults probably reaches much shallower depth.

The deformation observed on the Búrfellshraun profile can be interpreted as due to tensional failure in the region of the earthquake of December 5, 1968, some 10 to 15 kilometers southeast of the profile. This tensional failure has then produced a subsidence bowl around the epicenter, reaching as far as the Búrfellshraun profile. The displacement on fault B (Figure 5) is interpreted as a secondary effect of the earthquake, caused by plastic flow at depth towards the hypocenter, resulting in tensional failure at the surface. This failure may have occurred months after the earthquake. The uplift of the southeast part of the Búrfellshraun profile between 1969 and 1970 is interpreted as due to continuation of the plastic flow at depth which gradually has eliminated the stress anomaly which was caused by the earthquake of December, 1968.

CONCLUSION

The Reykjanes earthquake swarm of September, 1967 was caused by tensional failure along southwest to northeast trending fault or faults which extend towards the Vogar profile. The main effect of the earthquakes at the location of the Vogar profile was the creation of a zone of abnormally low pressure at depth. This zone of abnormally low pressure had a surface expression of a zone of subsidence. The low pressure at depth caused plastic flow in the lower part of the crust which reversed partly the surface effect of the earthquakes with a relaxation time of approximately one year. The earthquake in Southwest Iceland, December 5, 1968 was also caused by tensional failure, resulting in subsidence of the epicentral area. This subsidence was largely eliminated within 18 months after the earthquake.

A more detailed observation of the surface deformation that takes place during a period of one or two years after an earthquake should provide information on the thickness of the elastic crust and on the viscosity or plasticity of the lower part of the earth's crust.

REFERENCES

Einarsson, Trausti. The Icelandic fracture system and inferred crustal stress field, in Iceland and Mid-Ocean Ridges, ed. S. Björnsson, pp 128-139, Soc. Sci. Islandica, Publ. 38, 1967.

Haskell, N. A. Elastic displacements in the near-field of a propagating fault, Bull. Seismol. Soc. Amer., 59, pp 865-908, 1969.

Kjartansson, Gudmundur. The Búrfellshraun lava flow and its age, Náttúrufraedingurinn, 42, pp 59-183, 1973.

Klein, F. W., Páll Einarsson and M. Wyss. Microearthquakes on the Mid-Atlantic plate boundary on the Reykjanes Peninsula in Iceland, J. Geophys. Res., 78, pp 5084-5099, 1973.

Lawson, A. C., and others. The California Earthquake of April 18, 1906, Carnegie Institution of Washington, Publ. 87, Vol. 1, 451 pp, 1908.

Pálmason, Gudmundur. On heat flow in Iceland in relation to the Mid-Atlantic Ridge, in Iceland and Mid-Ocean Ridges, ed. S. Björnsson, pp 111-127, Soc. Sci. Islandica, Publ. 38, 1967.

Reid, H. F. The Mechanism of the Earthquake, Carnegie Institution of Washington, Publ. 87, Vol. 2, 192 pp, 1910.

Thoroddsen, Th. Die Geschichte der isländischen Vulkane, D. Kgl. Danske Vidensk. Selsk. Skrifter, Naturvidensk. og Mathem. Afd., 8 Raekke, IX, 458 pp, 1925.

Trifunac, M. D. A three-dimensional dislocation model for the San Fernando, California, earthquake of February 9, 1971, Bull. Seismol. Soc. Amer., 64, pp 149-172, 1974.

Tryggvason, Eysteinn. Fault displacement in recent Icelandic earthquakes, Trans. Am. Geophys. Union, 48, p 204, 1967.

Tryggvason, Eysteinn. Surface deformation and fault displacement associated with an earthquake swarm in Iceland, J. Geophys. Res., 75, pp 4407-4422, 1970.

Tryggvason, Eysteinn. Seismicity, earthquake swarms, and plate boundaries in the Iceland region, Bull. Seismol. Soc. Amer., 63, pp 1327-1348, 1973.

Tryggvason, Eysteinn, Sigurdur Thoroddsen and Sigurdur Thorarinsson. Report of earthquake risk in Iceland, Timarit Verkfraedingafélags Íslands, 43, pp 81-97, 1958.

Wyss, M., and J. N. Brune. Seismic moment, stress, and source dimensions for earthquakes in the California-Nevada region, J. Geophys. Res., 73, pp 4681-4694, 1968.

INDEX

GPSR Compliance

The European Union's (EU) General Product Safety Regulation (GPSR) is a set of rules that requires consumer products to be safe and our obligations to ensure this.

If you have any concerns about our products, you can contact us on ProductSafety@springernature.com

In case Publisher is established outside the EU, the EU authorized representative is:

Springer Nature Customer Service Center GmbH
Europaplatz 3
69115 Heidelberg, Germany

Batch number: 10761599

Printed by Printforce, the Netherlands